SPACE
EMERGING OPTIONS FOR NATIONAL POWER

DANA J. JOHNSON, SCOTT PACE, and C. BRYAN GABBARD

RAND • National Defense Research Institute

Approved for public release; distribution unlimited

The research was conducted in RAND's National Defense Research Institute, a federally funded research and development center supported by the Office of the Secretary of Defense, the Joint Staff, the unified commands, and the defense agencies under Contract No. DASW01-95-C-00059.

Library of Congress Cataloging-in-Publication Data

Johnson, Dana J.
   Space : emerging options for national power / Dana J. Johnson, Scott Pace, C. Bryan Gabbard.
      p.   cm.
   "Prepared by RAND's National Defense Research Institute."
   "MR-517."
   Includes bibliographical references.
   ISBN 0-8330-2493-0
   1. Astronautics, Military—United States. 2. Space industrialization—United States. 3. United States—Military policy. 4. World politics—1989 -   . I. Pace, Scott II. Gabbard, C. Bryan (Claybourne Bryan). III. National Defense Research Institute RAND (U.S.). IV. United States. V. Title.
UG1523.J64   1998
358' .8' 0973—dc21   97-12219
                                                                       CIP

RAND is a nonprofit institution that helps improve policy and decisionmaking through research and analysis. RAND's publications do not necessarily reflect the opinions or policies of its research sponsors.

© Copyright 1998 RAND

All rights reserved. No part of this book may be reproduced in any form by any electronic or mechanical means (including photocopying, recording, or information storage and retrieval) without permission in writing from RAND.

Published 1998 by RAND
1700 Main Street, P.O. Box 2138, Santa Monica, CA 90407-2138
1333 H St., N.W., Washington, D.C. 20005-4707
RAND URL: http://www.rand.org/
To order RAND documents or to obtain additional information, contact Distribution Services: Telephone: (310) 451-7002;
Fax: (310) 451-6915; Internet: order@rand.org

PREFACE

This report presents the results of a study conducted in 1993–1994 that examined the extent to which spacepower (both military and economic) will influence national security strategy and the conduct of future military operations. It was updated and revised in 1997 to reflect the changes that have occurred in military space policies, organizations, and operations and in the expanding commercial space sector since 1994. The report attempts to articulate the key military space policy issues facing the United States and place them in the larger context of a changing strategic environment to define new options for the exercise of spacepower in the pursuit of national interests.

The motivation for the study was twofold: (1) to educate decisionmakers on the exploitation of spacepower in the pursuit of national security interests, and (2) to provide an overview of economic security issues facing military planners who are already familiar with military space policies, programs, and trends. Since the research for the study was completed in 1994, many aspects of the trends discussed here concerning the commercial space market have come about. For example, a Presidential Directive on the Global Positioning System (GPS) was issued in March 1996, and a new National Space Policy was released in September 1996. Furthermore, several defense- and intelligence-community-related initiatives concerning space program management and operations that were begun in 1994–1995 are still unfolding as this document goes to print.

The research and analysis was conducted under the auspices of RAND's National Defense Research Institute, specifically the International Security and Defense Policy Center. NDRI is a federally funded research and development center sponsored by the Office of the Secretary of Defense, the Joint Staff, the unified commands, and the defense agencies.

The report should be of interest to persons concerned with U.S. military doctrine, strategy, policy, and force planning, as well as those interested in the future of civil/military relations in technology policy and space policy.

CONTENTS

Preface iii
Figures vii
Tables ix
Summary xi
Acknowledgments xvii
Acronyms xix

Chapter One
    INTRODUCTION 1
    Background 1
    Objectives and Approach 2
    Organization of This Document 3

Chapter Two
    THE "PROLIFERATION" OF SPACEPOWER: A GEOPOLITICAL
    AND POLICY CONTEXT 5
    Spacepower Defined 5
    The Proliferation of Conflicts, Threats, and Operations 8
    The Proliferation of Space System Capabilities to Meet an Expanding
        Variety of Conflicts 11
    The Proliferation of Actors in Space 17

Chapter Three
    TRENDS IN SPACE-RELATED FUNCTIONS: OPPORTUNITIES
    FOR COLLABORATION AND POSSIBILITIES FOR CONFLICT 21
    The Relationship Between Space-Related Functions and Sectors 21
    Space Launch: Government-Driven (in Transition) 22
    Satellite Communications: Commercially Driven 28
    Remote Sensing: Becoming Commercial 31
    Environmental Monitoring: Civil-Government-Driven 33

Satellite Navigation: Commercially Driven 35
Integrated Tactical Warning and Attack Assessment: Government-Driven 38
Space Control: Government-Driven 40
Force Application from Space/Ballistic Missile Defense: Government-Driven 42
Summary 44

Chapter Four
ILLUSTRATIVE MILITARY SPACE STRATEGY OPTIONS IN THE POST-COLD WAR WORLD 47
Illustrative Military Space Strategy Options 48
The Minimalist Strategy Option 50
The Enhanced Strategy Option 51
Aerospace Force Strategy Option 53
How the Options Accomplish Space-Related National Security Objectives 54
Accomplishing National Security Objectives: The Minimalist Strategy Option 56
Accomplishing National Security Objectives: The Enhanced Strategy Option 59
Accomplishing National Security Objectives: The Aerospace Force Strategy Option 61
How the Options Might Support Operations Across the Spectrum of Conflict 62
The Growing Importance of Trust and Space Literacy 65

Chapter Five
UPDATES SINCE 1994 67
Growth of Space Commerce 67
U.S. Policy and Strategy 68
U.S. Space Command 70
Works in Progress 73

Chapter Six
CHALLENGES FOR THE FUTURE 75
Choosing a Strategic Direction 81
Bottom Line 83

Bibliography 85

FIGURES

S.1.	An Organizational Perspective on the Three Strategies	xv
2.1.	Cold War Spectrum of Conflict	9
2.2.	A New Spectrum of Conflict	10
2.3.	The Proliferation of Actors Involved in Space Activities	18
2.4.	A "Hierarchy of Levels" in the U.S. Government's Space Policy Network	19
3.1.	Space-Related Functions on the Spectrum Are Driven by Different Sectors	22
3.2.	The Space Launch Market	23
3.3.	Percentage of Launches That Are DoD	26
4.1.	An Organizational Perspective on the Three Strategies	49

TABLES

2.1.	Illustration of Space Support for Peace, Peacekeeping/ Humanitarian, and Counterinsurgency Operations	14
2.2.	Illustration of Space Support for Anti-Terrorism, Crisis, and Country Conflict Operations	15
2.3.	Illustration of Space Support for Regional Conflict and Sustained Nuclear Operations	16
3.1.	Summary of Space-Related Functions and DoD Opportunities and Concerns	45
4.1.	How the Options Support Space-Related National Security Objectives	56

SUMMARY

INTRODUCTION

Operation Desert Storm clearly demonstrated that space forces can be significant contributors to enhancing operational forces and accomplishing military objectives. As a result, future joint operations will demand an increasing role for them. However, the strategic context in which many of these space forces were first developed, acquired, and justified has been supplanted by an evolving and dynamic landscape that poses new challenges for military forces in general. U.S. military space planners need to take a closer look at the operational implications of the choices they make when designing, acquiring, and operating space systems and forces. This document examines the changing nature of spacepower in the post-Cold War era in the conduct of military operations and the implications and challenges confronting policymakers in formulating future options for exercising spacepower.

THE "PROLIFERATION" OF SPACEPOWER

Before addressing the geopolitical and policy context for spacepower, the term itself must be defined. We would define spacepower as *the pursuit of national objectives through the medium of space and the use of space capabilities.* Although broad and general in nature, this definition focuses on national objectives, the use of space as a medium distinct from land, sea, or air, and the use of capabilities that require the space medium. The effective exercise of spacepower may require, but is not limited to, the use of military forces. We believe that "spacepower" should be viewed in a national context and that developing a strategy for spacepower must include consideration of economic and political security interests as well as military goals and objectives.

When we trace the role of military spacepower from the Cold War to the present, one clear pattern emerges: proliferation. We see this proliferation first in the increasing capabilities of space forces and the expanding roles they are expected to play to meet future missions and threats. During the Cold War, most military space systems were designed for strategic purposes (e.g., the deterrence of strategic nuclear conflict between the United States and the Soviet Union). Not only were they designed to provide warning of ballistic missile attack, they were also designed to communicate that warning through a nuclear environment and to support the U.S. nuclear response by helping to repel and defeat the attack and end the conflict on favorable terms.

In the post-Cold War era, the types of roles U.S. military forces may be called on to perform have increased dramatically and include peacekeeping and humanitarian operations, crisis operations, and theater defense. Besides continuing to perform the strategic deterrence function in the event of nuclear war, space forces are also expected to support U.S. responses to other forms of conflict by performing an array of space-related functions that include early warning and integrated tactical warning and attack assessment (ITW&AA),[1] weather/environmental monitoring, satellite communications (satcom), surveillance and reconnaissance, navigation and positioning, space control, and, possibly in the future, ballistic missile defense (BMD) and force application.

Beyond this proliferation of threats and space-related functions, there is also a proliferation in the number of players in the space arena. In the past, the U.S. military overwhelmingly dominated spacepower. Now, however, a number of actors in several sectors are involved in space programs and activities and have an effect on national security: the military space sector, the intelligence space sector, the commercial space sector, and the civil space sector. In addition, an international sector has a strong influence on the activities of these other sectors, which are becoming much more intertwined and interconnected, increasing the complexity involved in military exploitation of spacepower.

TRENDS IN SPACE-RELATED FUNCTIONS

Given that the military no longer dominates the market for space-derived information, it is critical that military policymakers understand the economic and commercial trends in the various functional areas where space forces will be expected to provide support, since those trends affect the development, acquisition, deployment, and exploitation of space systems, posing risks and offering opportunities. Space-related functions can be arrayed along a spectrum, with commercial, civil, intelligence, international, and military sectors having differing degrees of importance in each. Underlying the spectrum is the

---

[1]ITW&AA consists of two separate functions: tactical warning and attack assessment. Although they can be performed separately, integration of the two (the "I" in "ITW&AA") implies the need to integrate the functions to have more information with which to determine an appropriate response to a ballistic missile attack.

function of space launch, which is a prerequisite for all aspects of spacepower. U.S. dominance in this area fell dramatically through the 1980s. To ensure better access to space, and to gain additional military and diplomatic leverage, a reliable, responsive, lower-cost space launcher is needed.

At the commercial end of the spectrum is the satellite communications function, which is driven by commercial interests in terms of number of customers, of money, and, increasingly, of deploying new technologies. DoD use of commercial communication satellites can bring technical and cost benefits, but at the risk of additional vulnerability and inflexibility compared to completely U.S.-government-owned systems. With its origins in the military, space-based remote sensing is increasingly becoming commercial, as technical advances by DoD and the Department of Energy, bolstered by policy changes, open up the prospect of smaller, lighter, more selective, and less-expensive remote sensing. While there are economic benefits to the military (in terms of procuring commercially available imagery), commercial remote sensing has its downside—it poses the risk of possibly revealing the disposition and movement of U.S. forces to an adversary. Like remote sensing, satellite navigation is a function that satisfies a long-standing critical military need and that is increasingly becoming of greater commercial interest as the opportunities for new geographically based information systems become more widely known. The DoD operates a constellation of 24 Navstar satellites that make up the space segment of the Global Positioning System (GPS). DoD control over the space segment protects U.S. military interests but entails the responsibility of maintaining a predictable and stable policy environment for GPS to ensure that it becomes a global standard, to deter the proliferation of competing systems, and to allow U.S. industry to maintain its leadership position in growing commercial markets.

In the civil part of the spectrum, environmental monitoring poses concerns for DoD, since until recently both DoD and the National Oceanic and Atmospheric Administration (NOAA) have maintained separate weather satellite programs. This overlap causes problems over issues of encryption and over the use of declassified data from military systems in environmental research.

At the military end of the spectrum, the military controls such functions as ITW&AA, space control, and BMD. These functions represent capabilities that perform critical national-security-related missions, such as ensuring the survival and protection of the United States from threat of attack, which, therefore, probably will not be contracted out to the commercial sector. Military control of these functions affords the United States opportunities for exercising post-Cold War leadership in support of the stable environment needed for global economic growth and as a counterbalance to aggressive regional powers.

ILLUSTRATIVE MILITARY SPACE STRATEGY OPTIONS IN THE POST-COLD WAR WORLD

To understand the implications of operating in the post-Cold War environment, we look at three illustrative evolutionary military space strategy options that span the range of potential military involvement in space-related functions: Minimalist, Enhanced, and Aerospace Force. In the Minimalist option, the military use of spacepower is highly dependent on external relationships and partnerships. Integration with other military operations depends on organizations outside the military chain of command. This strategy option is largely the outcome of budgetary constraints and technological advances in other sectors, thus leading to the U.S. military owning only those systems that perform unique and/or critical national security functions and leasing everything else from the commercial sector. In the Enhanced strategy option, the military use of spacepower is highly integrated with other forms of military power. External relationships and partnerships are important but not critical to core military capabilities. In the Aerospace Force option, military spacepower is exercised separately from other military forces. Actual military operations are most likely joint and combined and may use external relationships, but this is not required. Figure S.1 shows the organizational implications of the options.

We also considered how each option would accomplish space-related national security objectives, including: (1) preserving freedom of, access to, and use of space; (2) maintaining the U.S. economic, political, military, and technological position; (3) deterring/defeating threats to U.S. interests; (4) preventing the spread of weapons of mass destruction to space; and (5) enhancing global partnerships with other spacefaring nations. Although all the options can fulfill the objectives, how they fulfill them differs. For example, the Minimalist option needs to rely on multiple means, such as economic strategies, terrestrial military forces, diplomatic approaches, and treaties, to accomplish its objective, since it has few actual space forces capable of doing so. We also examined how each option might support operations across the spectrum of conflict, and found that each option will need very different organizational constructs to fulfill the tasks and functions required. For example, a Minimalist option for DoD exploitation of space-based systems—a notion that is entirely conceivable in today's budgetary environment—necessitates much earlier "preparation of the battlefield" through possibly years of expanded DoD and military involvement in such nontraditional areas as trade policies and regulations. However, regardless of option, there is a growing need for improved cooperation among all sectors. This, in turn, entails a growing need for space literacy about what space forces can and cannot do, and for trust among the various players, both within the services and across the various sectors.

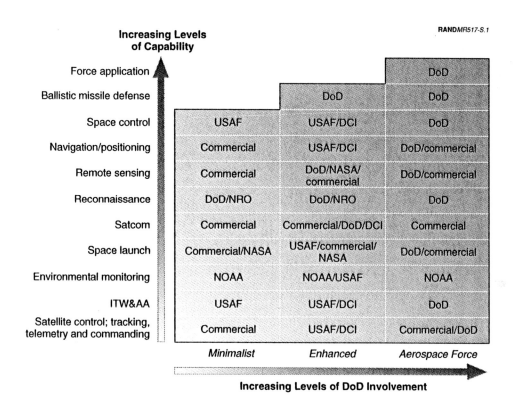

Figure S.1—An Organizational Perspective on the Three Strategies

POSTSCRIPT

Since this report was first written, a number of changes in national policies, organizations, operations, and commercial sector trends have occurred, having a bearing on the development and exploitation of spacepower. At the policy level, the release of the National Security Policy, a new National Space Policy, and *Joint Vision 2010* influences the development of requirements that spacepower will be expected to support. USSPACECOM initiated several efforts that further defined these requirements by conducting a revalidation of its mission and organization in 1996. Furthermore, the focus of space warfighting has shifted from a strategic, Cold War orientation to a greater emphasis on the tactical exploitation of space and support to the warfighter in conventional operations. The influence of the "revolution in military affairs" (RMA) and the attention paid throughout the government to "information operations" and "information warfare" also point to the contribution of space-based systems and information technologies to supporting these areas of potential operational and institutional transformation. However, this new level of thinking about the military uses of space highlights the need for the military space community to come to grips with how to apply existing policy and doctrine to the specific challenges of using spacepower. This is admittedly a difficult problem, but the challenges posed by the expanding use of the medium of space by military and nonmilitary actors warrant a greater effort by the DoD to address how the commercial and civil space sectors can contribute to accomplishing national security goals.

CHALLENGES FOR THE FUTURE

Finally, U.S. spacepower comprises national capabilities, not just the military space capabilities of the DoD. The continued growth of the commercial space sector, especially in information technologies, is providing new opportunities for the exercise of national spacepower for military, political, and economic objectives. Regardless of the outcomes of roles and missions debates, reorganization and management initiatives, or defense force structure reviews, the warfighter will need to have a better understanding of how to exploit and counter commercial space capabilities. This understanding will better enable the military to create mechanisms, whether legal, regulatory, or market-driven, to ensure that it can "shape the battlefield" of the future and gain maximum leverage from whatever resources it is able to devote to military space capabilities. The DoD has begun the process of improving interoperability between military, civil, and commercial systems, and of ensuring it has access to commercial space systems when needed. But a deliberate effort at coordination and communication between representatives of the nation's military, economic, and political interests in space, built on a strong foundation of trust, literacy, and cooperation, is critical. Only then will we understand the extent to which spacepower will influence the implementation of national security strategy and the conduct of future military operations in the context of exercising national power in a dynamic strategic environment.

ACKNOWLEDGMENTS

We could not have undertaken this study without extensive support from many individuals and organizations throughout the U.S. national security and space communities. The authors are grateful for the guidance and assistance provided by Colonel Roger Graves, USAF, United States Space Command/J5; Colonel Robi Chadbourne, USSPACECOM/J5; Colonel Ken Rosebush, USAF, Air Force Space Command/XPX; and Colonel James Dill, USAF, Onizuka Air Force Base, California. Also at USSPACECOM/J5, Colonel William Mulcahy, USAF; Commander Terry Dorphinghaus, USN; Lieutenant Colonel Rudy Veit, USA; Major Dave Girard, USA; and others were very helpful in providing their insights and guidance. The authors also benefited from discussions held with the staffs of two other offices within USSPACECOM—J35, headed by Colonel Steve Sloboda, USAF, and J33S. At the Air Force Space Command, Brigadier General Roger DeKok, Director of Plans (XP), gave freely of his time and advice; in addition, Lieutenant Colonels Michael Wolfert, Jon Noetzel, and Randy Joslin, and Majors Sam Lee, Steve Prebeck, and Kurt Stevens were particularly helpful in providing their insights into command thinking about future space forces, launch systems, and evolving concepts such as information warfare. Major Patrick Rayermann of the Army Space Command was also very helpful in providing his command's perspective on space support to the warfighter and other issues.

The authors are deeply indebted to Colonel Jim Burke, USAF, in the Pentagon, for his long-term support and guidance. Others in the Pentagon who provided extensive commentary and insights into the space policy process include Colonel Simon "Pete" Worden, USAF; Colonel Gilbert Siegert, USAF; Lieutenant Colonel Michael Brown, USA; Lieutenant Colonel Mark Rochlin, USA; Lieutenant Colonel Bob Work, USMC; Major Bob Butler, USAF; Mark Berkowitz; and Jordan Katz. The former Deputy Secretary of Commerce, Thomas Murrin, and Courtney Stadd from the National Space Council enlightened us on the intricacies of interagency negotiations. At the Office of Science and Technology

Policy, Richard DalBello and Jeff Hoffgard gave us valuable insights into current administration thinking on space policies and programs.

In the private sector, we had numerous fruitful discussions with Charles Trimble and Ann Ciganer (Trimble Navigation), Dr. Jack Oslund and his colleagues at COMSAT, James Frelk (Lockheed), Wallace McClure and Jeffrey Morrow (Rockwell International), and others.

We also benefited from lengthy discussions with faculty and staff participating in Spacecast 2020 at the Air War College, Maxwell AFB, in October 1993.

The authors are indebted to their RAND colleagues Bruno Augenstein, David Chu, Glenn Buchan, Elwyn Harris, Katherine Poehlmann, and Gaylord Huth for their insights, suggestions, and guidance. Glenn Buchan and Kevin O'Connell provided critical observations in their reviews of this document, which were much appreciated. Betty Ashford, Rosalie Fonoroff, Birthe Wenzel, and Patricia Bedrosian gave the authors extensive support in the management and production of this study. Finally, Paul Steinberg gave us insightful observations, crucial recommendations, and encouraging support throughout the process of completing this report.

To all of these people we are indebted, and any errors, oversights, and statements of opinion are those of the authors alone.

ACRONYMS

AFSATCOM	Air Force Satellite Communications System
AMC	Air Mobility Command
AOR	Area of Responsibility
ARPA	Advanced Research Projects Agency
ASAT	Antisatellite
AWACS	Airborne Warning and Control System
BMD	Ballistic Missile Defense
BMDO	Ballistic Missile Defense Organization
CAIV	Cost as an Independent Variable
CALC	Crises and Lesser Conflicts
CINC	Commander-in-Chief
CINCSPACE	Commander-in-Chief, United States Space Command
COMAFSPACE	Commander of AFSPACE Forces
Comsat	Communications Satellite
CONUS	Continental United States
CRAF	Civil Reserve Air Fleet
CSCI	Commercial Satellite Communications Initiative
DCI	Director of Central Intelligence
DGPS	Differential GPS
DISA	Defense Information Systems Agency
DMSP	Defense Meteorological Satellite Program
DoC	Department of Commerce

DoD	Department of Defense
DoE	Department of Energy
DoI	Department of the Interior
DoT	Department of Transportation
DSCS	Defense Satellite Communications System
DSP	Defense Support Program
ESA	European Space Agency
FAA	Federal Aviation Administration
GCCS	Global Command and Control System
GN&C	Guidance, Navigation, and Control
GPALS	Global Protection Against Limited Strikes
GPS	Global Positioning System
HRMSI	High Resolution Multispectral Imaging Sensor
ICBM	Intercontinental Ballistic Missile
IRBM	Intermediate Range Ballistic Missile
ITW&AA	Integrated Tactical Warning and Attack Assessment
JFACC	Joint Force Air Component Commander
JSTARS	Joint Surveillance Target Attack Radar System
LEO	Low Earth Orbit
MILSTAR	Military Strategic and Tactical Relay (Satcom System)
MRC	Major Regional Conflict
MTCR	Missile Technology Control Regime
NASA	National Aeronautics and Space Administration
NAVWAR	Navigation Warfare
NCA	National Command Authority
NGO	Nongovernmental Organization
NOAA	National Oceanic and Atmospheric Administration
NORAD	North American Aerospace Defense Command
NRO	National Reconnaissance Office

NSC	National Security Council
POES	Polar Orbiting Environmental Satellite
R&D	Research and Development
RMA	Revolution in Military Affairs
satcom	Satellite Communications
SBIRS	Space-Based Infrared System
SDI	Strategic Defense Initiative
SecDef	Secretary of Defense
SOC	Space Operations Center
TBM	Theater Ballistic Missile
TENCAP	Tactical Exploitation of National Capabilities
TMD	Theater Missile Defense
TT&C	Telemetry, Tracking, and Control
UCP	Unified Command Plan
UFO	Ultra-High Frequency Follow-On (Satcom System)
USAF	United States Air Force
USSPACECOM	United States Space Command
USTRANSCOM	U.S. Transportation Command
WMD	Weapons of Mass Destruction
WRC	World Radiocommunications Conference

Chapter One
INTRODUCTION

BACKGROUND

Operation Desert Storm clearly demonstrated that space forces can be significant contributors to enhancing operational forces and accomplishing military objectives. They provide critical data and information necessary to warn against ballistic missile attack, to allow instantaneous worldwide communications among forces, to predict weather patterns in regions of national interest, and to perform precise geographical measurements and position location anywhere in the world. Coupled with advances in technologies and information processing techniques, space systems are demonstrating their flexibility and diversity in other areas as well, such as contributing to the development of national communication infrastructures in developing countries. The satellites used to perform these many functions are only part of a larger system that allows the United States to exploit space-based capabilities to enhance its national security; the components of this system include launch vehicles to ensure access to space, ground stations deployed around the globe to track and control the satellites, and reliable communication links to ensure the timely flow of data to and from the orbiting spacecraft. Also included is the large infrastructure of unique skills and technologies tailored to exploit and disseminate the products of space resources.

Although Operation Desert Storm highlighted the importance of spacepower in supporting conventional military operations, future joint operations will demand an increasing role for space forces, including the possible exploitation of civil, commercial, and international space systems. Warfighters, users, and military planners must also be concerned with the implications for future military operations of facing adversaries who have significant space capabilities or who have access to space-derived data products. Moreover, the implications of such future concepts as information warfare need to be addressed. In this dynamic security and budgetary environment, members of the military space community must determine even more carefully which space forces and operational arrangements are needed, when they are needed and by whom, and what the operational implications are of the space force structure that results.

The strategic context in which many of these space systems were first developed, acquired, and justified has been supplanted by an evolving and dynamic strategic landscape that poses new challenges for military forces in general. Consequently, space forces must now compete with other military systems and capabilities in a declining budgetary and force structure environment. It is not evident that a clear understanding of the virtues and drawbacks of space assets exists among key decisionmakers in the administration, the Congress, or the American public. This lack of understanding is complicated by a gap of perspective and experience between those in the military and the aerospace industry tasked to develop, acquire, and operate space-based capabilities (the space community) and those in the combatant and component commands who plan and conduct combat operations to which these capabilities might contribute (the warfighting community). This gap has hampered the efficient communication needed to fully exploit existing capabilities and to remedy deficiencies by effectively upgrading or replacing associated systems. Other analysts have noted that if the two communities cannot readily and effectively communicate with each other, they surely will not be able to communicate with those civilian leaders in the Pentagon, the White House, and on Capitol Hill responsible for making informed, fiscally responsible decisions on acquiring capabilities—the all-important connection between space-based capabilities and national objectives.

U.S. military space planners have begun taking a closer look at the operational implications of the choices they make when designing, acquiring, and operating their systems. In particular, efforts are under way to close the general schism between the space and warfighting communities. Still, much effort is needed to ensure that all decisionmakers are operating from a common frame of reference.

OBJECTIVES AND APPROACH

This report presents the results of a 1993–1994 study to help in establishing that frame of reference. Specifically, the study examines the changing nature of spacepower in the post-Cold War era in conducting military operations and the implications and challenges policymakers face in formulating future options for exercising spacepower. Since many changes in policy, doctrine, organization, and commercial space activities have occurred since 1994, the study brings those areas up to date and addresses their implications for spacepower.

The motivation for the study was twofold: (1) to educate decisionmakers on the exploitation of spacepower in the pursuit of national security goals and objectives, and (2) to provide an overview of economic security issues facing military planners who are already familiar with military space policies, programs, and trends. Since the completion of the research for this study in 1994, many aspects of the trends discussed in this document have occurred. For example, a number of Presidential decisions regarding GPS and national space policy have been promulgated, and efforts toward reorganizing the management of DoD and intelligence community space programs have been initiated. The implications of these steps are still unfolding as this document goes to print.

The alternative options discussed herein are intended to be plausible but illustrative options that could evolve from the status quo. They are created to help illuminate the issues and challenges for spacepower in supporting national security objectives and, thus, are not rigorously evaluated.

ORGANIZATION OF THIS DOCUMENT

Chapter Two traces the evolution of spacepower and space forces from the Cold War to present times, highlighting the proliferation in military threats, the proliferation of space-based capabilities to address those threats, and the proliferation of actors involved with space systems. Chapter Three discusses the issues and trends underlying this proliferation, examining the status of the various players involved in the pertinent areas of space activity, the areas that offer opportunities for mutually beneficial cooperation between military and civil/commercial sectors, and the areas that might cause concerns for competition. Given the changing environment highlighted in Chapter Two and the trends and issues identified in Chapter Three, we examine in Chapter Four the relationship of spacepower and the evolving national security environment. The chapter addresses alternative interpretations of spacepower for different futures by creating an illustrative spectrum of options for exercising space force, examining the implications of those options for meeting space-related national security objectives, and assessing how they might accomplish operational objectives and tasks. Chapter Five offers a postscript of what "vision," policy, and organizational changes have occurred since the initial completion of this study. These changes include the promulgation of *Joint Vision 2010* by the Staff and USSPACECOM's own vision of the future. They also include an examination of the Air Force's vision of evolving from today's "air" dominated air force, to an "air and space" force, and eventually to a future "space and air" force. A new National Space Policy has also been released (in September 1996). Finally, Chapter Six summarizes the importance of economic and commercial space interests for the military, arguing that cooperation and trust among all sectors will enable everyone to better understand the extent to which spacepower can influence the exercise of national power in a dynamic security environment.

Chapter Two
THE "PROLIFERATION" OF SPACEPOWER: A GEOPOLITICAL AND POLICY CONTEXT

When we trace the role of military space forces from the Cold War to the present, one clear pattern emerges: proliferation. We see this proliferation in the increasing capabilities of space forces and the expanding roles they are expected to play to meet future missions and threats. We also see this proliferation in the expanding number of actors currently involved—the number of "players" who have a stake in using space forces. Understanding this context is critical for understanding how the military should fashion its space strategy.

This chapter establishes the geopolitical and policy context for identifying this proliferation of purposes and users of space-derived information. It starts by defining the term "spacepower," which provides the context for the theme underlying this report, that spacepower should be viewed in a national framework and that developing a military strategy for spacepower must include consideration of economic and political security interests as well as military goals and objectives. The chapter continues by examining the dynamic and evolving spectrum of conflicts and threats and the operational tasks that space systems would be expected to support across that spectrum. It then discusses the different sectors of space activities (military, intelligence, civil, commercial, and international) and the existing policy process and key players in that process.

SPACEPOWER DEFINED

What is spacepower? In an analogy to air and sea power, the term would seem to imply the employment of military forces operating in a distinct medium (i.e., the space environment) to achieve some national goal or military objective. Current Air Force doctrine defines *spacepower* as the "capability to exploit space forces to support national security strategy and achieve national security objectives."[1] It also defines *air and space power* as "the synergistic application

---
[1] Air Force Basic Doctrine (1997).

of air, space, and information systems to project global strategic military power." Unfortunately, these definitions seem to be incomplete; they do not capture some important realities of military space activities today and in the future.

First, there is the implied assumption that the identification of military space forces provides both the necessary and sufficient conditions for understanding the space "order of battle." As will be discussed below, civil and commercial systems are also an important part of the nation's space capabilities and its ability to achieve national security objectives. Partnerships between military, civil, and commercial communities are likely to be vital to the successful execution of national and military security strategies. Thus, spacepower should be understood as more than military forces. As General Hap Arnold said of airpower: "Air power is the total aviation activity—civilian and military, commercial and private, potential as well as existing." We would apply the same thought to the definition of spacepower.

Second, the definitions imply that spacepower is focused on "global" and "strategic" concerns alone. This is understandable, as space forces have historically been thought of as systems for functions such as strategic nuclear operations and national intelligence collection. It is, however, an overly narrow and outmoded view. *Joint Vision 2010* defines four new operational concepts for the U.S. military: dominant maneuver, precision engagement, full-dimensional protection, and focused logistics. Space systems contribute to and are necessary to each of these operational concepts. Space capabilities are also vital to military operations other than war, as in the case of peacekeeping and humanitarian relief. Consequently, space forces are more than a tool for achieving global, strategic objectives as during the Cold War—they are an integral part of how U.S. forces will operate across the spectrum of conflict.

Third, the definitions give an impression of being taken at one point in time, e.g., at the instant during which power is being projected in support of a national objective. Power can be thought of not only as the ability to employ forces, but also as the ability to shape the battlespace itself *before the initiation of conflict*. As with other forms of national power, both absolute and relative capabilities are important, e.g., What are my forces capable of doing and how do they compare to those of potential adversaries? In line with the National Military Strategy, spacepower can be used to shape and mitigate potential threats and not just respond to them.[2] Since spacepower is more than military forces alone, this means that spacepower should be understood as something

---

[2]Office of the Joint Chiefs of Staff (1997).

that can evolve and that the ability to shape the actions of others may be as significant as what can be accomplished unilaterally.[3]

As with any evolving military field, intense debates over doctrine can be expected. Like the emergence of air- and seapower, spacepower has both similarities and differences with other forms of military and national power. Spacepower has many different facets depending on one's perspective and objectives. For example, from the viewpoint of the tactical commander, spacepower represents capabilities that can help him put bombs on target in his area of the battlespace:

From the viewpoint of the regional CINC, spacepower represents capabilities that shape the entire battlespace, including the provision of logistical support and the use of joint and combined arms capabilities. His view is a broader than that of the lower-level commander:

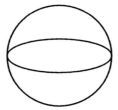

From the viewpoint of the National Command Authority, the battlespace is but one part of several areas of concern, such as domestic political support, relations with allies and coalition partners, and economic conditions. The viewpoint at this level can be thought of as monitoring several different but interacting spheres:

---

[3]For example, having low-cost access to space is useful by itself and as an additional deterrent to the entry of potential competitors, which may result in missile proliferation. Similarly, the continued provision of free, high-quality navigation signals from GPS makes it difficult to raise international funding for a competing system.

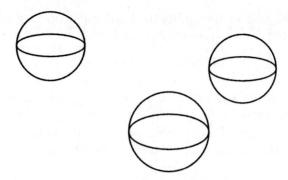

Spacepower is connected to other forms of national power such as economic strength, scientific capabilities, and international leadership. The NCA may use spacepower to achieve nonmilitary objectives or exploit nonmilitary capabilities to enhance military spacepower.

An examination of spacepower should consider all of the nation's space capabilities, at all levels of conflict, and across time to include shaping the battlespace in peacetime before the initiation of conflict. Therefore, we would define spacepower as *the pursuit of national objectives through the medium of space and the use of space capabilities.* Although broad and general in nature, this definition focuses on national objectives, the use of space as a medium distinct from land, sea, or air, and the use of capabilities that require the space medium. The effective exercise of spacepower may require, but is not limited to, the use of military forces.

This report focuses on the use of spacepower at the strategic and operational levels of warfare, e.g., the needs of the CINC, rather than those of the tactical commander. The exercise of spacepower by tactical commanders requires a more technical and detailed analysis of specific space capabilities. This report provides an overview of how the exploitation of space is creating new opportunities for the exercise of national power.

THE PROLIFERATION OF CONFLICTS, THREATS, AND OPERATIONS

Figure 2.1 shows the conflict spectrum during the heyday of the Cold War. At the time, American foreign policy was determined by the threat posed by the Soviet Union and its surrogates in the Warsaw Pact. Military planners focused on strategies for deterrence and defense: deterring a nuclear conflict with the USSR but defeating it should an attack be initiated. The likelihood of such a conflict was believed to be low, but the consequences of such a catastrophe warranted extensive preparation to ensure the government's survival and effective retaliation. In contrast, less emphasis was placed on conventional

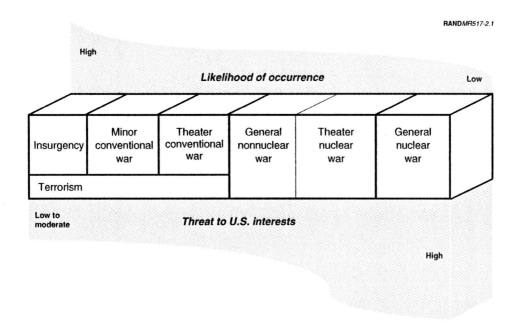

Figure 2.1—Cold War Spectrum of Conflict

operations and capabilities to deal with insurgencies, terrorism, and smaller conventional conflicts, because the threat to the United States itself was perceived to be low and national interests might or might not be at stake.

However, the end of the Cold War has brought a dynamic international environment that has engendered new uncertainties and concerns about the diversity of threats to U.S. military and economic interests and the extent to which the United States will become involved in military operations overseas (as shown in Figure 2.2). Although the possibility of a theater or general nuclear war between the United States and Russia or other nuclear-armed states cannot be discounted (these states still possess intercontinental ballistic missiles that could threaten U.S. national survival), it has certainly diminished (and is perceived to be low); however, other threats have taken its place. Proliferation of weapons of mass destruction (WMD) in other regions of the world, possibly including space, contribute to the uncertainties that military planners must consider. Operational planning must account for the possibility of these threats while recognizing that the United States is more likely to become involved in major regional conflicts (MRCs) similar to Desert Storm, as well as in crises and

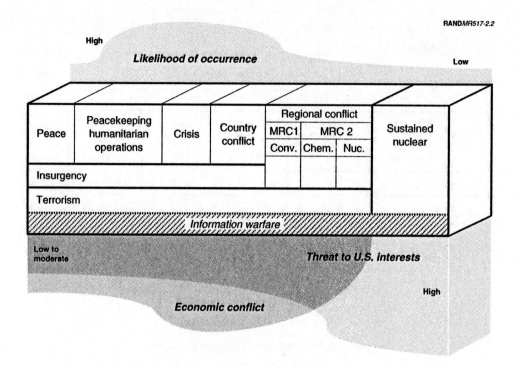

Figure 2.2—A New Spectrum of Conflict

lesser conflicts (CALCs)[4] such as peacekeeping, rescue, and humanitarian operations. "Peace" in this instance is a relative term, for although the threat to the United States may be low in these cases, they increasingly require a greater commitment of military forces and a concomitant risk of escalation to greater combat involvement.

There are currently more than 70 flashpoints worldwide.[5] Reflecting these flashpoints, there are an expanded number of global actors with which the United States must be concerned. These include transnational threats such as international narcotics and other criminal activities that are well-funded, have

---

[4]CALC is a RAND-coined term used to describe international situations involving nonroutine military operations short of war or short of preparations for war. More broadly, it includes everything that is not an MRC. Examples of CALCs include peacekeeping and peacemaking (Lebanon, Somalia), punitive and preemptive strikes (Libya, Iraq), restoring civil order (Somalia, Grenada, Panama), humanitarian and disaster assistance (Bosnia, Somalia), evacuations of American citizens (Liberia, Somalia), provision of security zones (Iraq, Bosnia), and monitoring and enforcement of sanctions (Iraq, Serbia). It does not include military support in domestic natural disasters, such as Hurricane Andrew. See Builder and Karasik (1995).

[5]Fogleman (1994), p. 7B. At the time of this article General Fogleman was the Commander-in-Chief, U.S. Transportation Command (USTRANSCOM), and Commander, USAF Air Mobility Command (AMC).

access to very sophisticated technologies and systems, and are growing in sophistication.[6] They also include nongovernmental organizations (NGOs), small terrorist groups, and insurgents. Separatist and ethnic conflicts continue to cause death and destruction in many parts of the world. The traditional problem of containing expansionist powers has been replaced in many regions by the opposite problem of supporting weak or collapsed governments (e.g., Somalia, Haiti, the former Yugoslavia). Thus, although the national security community worries about whether it can meet the demands of two major regional contingencies, such as wars in Iraq and Korea, the U.S. military finds itself actually engaged in a range of multiple, small, but difficult operations that are straining key support functions (e.g., airlift, medical services, communications).[7] Economic competition and economic conflict play a role across the board. Intensive economic competition can exist between two or more nations in a period of peace, and can contribute to shaping the environment in which military forces may operate. Alternatively, economic conflict may exist, because of sanctions and other measures, with little or no economic competition at higher levels of conflict. The disruption of information systems or the exploitation of information-based technologies by adversaries could also conceivably occur across the spectrum of conflict—and could, in turn, affect the U.S. military's ability to support national objectives at whatever level.

This proliferation of threats will likely result in asymmetrical relationships and confrontations in which many of the U.S. military's traditional doctrinal approaches to warfare and the capabilities it relies on may be inappropriate. The recent American experience in Somalia comes to mind as an example.

THE PROLIFERATION OF SPACE SYSTEM CAPABILITIES TO MEET AN EXPANDING VARIETY OF CONFLICTS

During the Cold War, many military space systems, such as the Defense Support Program (DSP) and MILSTAR, were designed for strategic purposes (i.e., the deterrence of strategic nuclear conflict between the United States and the Soviet Union). Not only were they designed to provide warning of ballistic missile attack (not only from the Soviets, but worldwide), they were also designed to communicate that warning through a nuclear environment and to support the U.S. nuclear response by playing a role in helping to repel and defeat the attack and end the conflict on favorable terms. These (and other) strategic purposes provided the context for establishing heroic system survivability criteria, which greatly increased the cost of many systems. Nevertheless,

---

[6]The Russian "mafia" is a source of particular concern with the collapse of internal controls in the former Soviet Union. See Williams (1994).

[7]D. Johnson (1994).

these systems contributed to ensuring deterrence and maintaining the nuclear peace during the Cold War.

Since Operations Desert Shield and Storm, however, military space systems are playing an increasingly critical role in enhancing the performance of American military forces across the board. Besides continuing to provide strategic deterrence, space forces today are intended to play a role in fulfilling a military strategy of near-simultaneous operation and support of two MRCs and a role in supporting the host of nontraditional missions exemplified by CALCs. MRCs tend to be technology-intensive, whereas CALCs are manpower-intensive. The contributions of space systems to meeting the demands of Gulf War operations have been widely heralded. However, the stresses imposed on space systems in providing that support have also been documented.[8] Although many of the problems identified in the Gulf War post-mortems have been addressed, there is a growing concern that the United States may become unable to support the two-MRC strategy if the level of forces (including space forces) is allowed to continue declining for the next decade. One way to deal with this problem of growing commitments and declining forces and budgets would be to alter the two-MRC strategy itself to reflect political and budgetary reality, as has been discussed in various public fora.

Furthermore, it is likely that the types of space forces, personnel, and support assets involved in CALCs will be stressed severely, in no small measure as a result of the continued drawdown in forces coupled with increasing international and domestic pressures for U.S. involvement. In most, if not all, cases for both MRCs and CALCs, the United States tends to be—or is expected to be—the dominant partner in these coalitions, with sometimes great disparities in levels of technologies, interoperability, training, exercises, and operational experience among the coalition partners. This places the burden on the U.S. military to provide operational support to its coalition partners, in addition to supporting its own forces, which is itself difficult in times of fiscal austerity. As the time for replacing many current satellite constellations draws near, the inability of the United States to pay fully for replacements will force the creation of innovative approaches to space system acquisition, operation, and employment to meet national commitments.

Replacement systems need not be as expensive as current systems and, in many cases, can perform better while costing less. Some acquisition reform efforts by the Air Force are demonstrating the kinds of innovative approaches that can bridge the gap between evolving requirements and available budgets. For example, the Global Positioning System is the largest operational military satellite

---

[8]See, for example, Winnefeld, Niblack, and Johnson (1994), especially Chapter Eight.

constellation to date. By using performance-based specifications and best commercial practices, the GPS program office reduced the cycle time for the next block of GPS sustainment satellites (the Block IIF series) from seven to five years, saved $1.1 billion over the acquisition, and reduced required project office manpower by 38 percent (i.e., from 145 persons in FY 96 to 90 by FY 08).[9]

In another example of acquisition innovation, the new Space-Based Infrared System (SBIRS) is intended to replace the current Defense Support Program (DSP) for such purposes as missile warning, missile defense, technical intelligence, and battlespace characterization. The "high" portion of SBIRS consists of four satellites in geosynchronous orbit and two satellites in highly elliptical orbits. Despite higher performance requirements, total life-cycle costs are estimated to be less than DSP because of smaller launch vehicles (e.g., the Medium Launch Vehicle as opposed to the Titan IV), the use of a commercial spacecraft bus, and the use of techniques such as CAIV (cost as an independent variable) to determine what the "best value" approach is for meeting user requirements. Continuation of acquisition reform and its "institutionalization" will be a necessary part of executing the National Security Strategy and National Military Strategy in a constrained budget environment.

The spectrum of conflict in Figure 2.2 showed the wide range of operations that the U.S. military may be called on to support today: peacekeeping/ humanitarian operations, insurgency, terrorism, crisis, country conflict, regional conflict (MRC level), and sustained nuclear operations. Furthermore, it is likely that the military will contribute to supporting whatever actions U.S. decisionmakers find necessary in the realm of information warfare.[10] Tables 2.1 to 2.3 show how a series of space-related functions could contribute to that range of operations. Those functions are integrated tactical warning and attack assessment (ITW&AA), weather/environmental monitoring, satellite communications (satcom), surveillance and reconnaissance, navigation and positioning, space control, ballistic missile defense (BMD), and force application. For each type of operation, we include an overall objective for the operation and specific tasks for each function. Rather than attempting to be all-inclusive—an impossible task, given the wide variety of uses of space-derived information—the tables are intended to illustrate the range of support that space systems provide for the accomplishment of objectives and tasks.

---

[9]Assistant Secretary of the Air Force (Acquisition), http://www/safaq.af.mil, 26 November 1996.

[10]See, for example, Molander, Riddile, and Wilson (1996); Anderson and Hearn, RAND, personal communication. Also, on the list of joint doctrine publications put out by the Joint Staff is Joint Publication 3-13, *Information Warfare*, which, as of the fall of 1996, was under development.

14   Space: Emerging Options for National Power

Table 2.1
Illustration of Space Support for Peace, Peacekeeping/Humanitarian, and Counterinsurgency Operations

Type of Operation and Operational Objectives	ITW&AA	Weather/ Environmental Monitoring	Space Launch	Satcom	Surveillance/ Reconnaissance	Navigation/ Positioning	Space Control	BMD	Force Application
Peacetime Operations									
Inhibit surprise attack	Detect, track, and assess missile launches		Provide capability to deploy satellites on short notice				Ensure survivability of space assets	Deter threat of ballistic missile attack	Deter threat of attack against U.S. forces overseas
Monitor arms control agreements				Support U.S. diplomatic operations	Monitor treaty compliance				
Provide routine operational support		Monitor weather for military exercises	Maintain adequate satellite constellations	Maintain secure communications to forces		Provide position location to troops in field	Routinely monitor satellites in orbit		
Peacekeeping/ Humanitarian									
Establish and defend safe areas	Monitor theater ballistic missile launches in areas of crisis			Ensure secure communications between U.S. and coalition forces	Deny infiltration in regions of concern	Establish accurate boundaries of safe areas	Ensure freedom of movement to, from, and in space	Monitor potential threats to safe areas	
Protect and rescue U.S. citizens overseas		Assess terrain for rescue operations			Monitor aggressive troop movements				Deter threat of attack against U.S. forces engaged in hostage rescue
Conduct disaster relief		Monitor weather for humanitarian operations	Maintain adequate satellite constellations	Establish communication links in disaster areas	Assess terrain for rescue operations				
Counterinsurgency									
Provide foreign internal defense support		Monitor weather over areas of interest		Ensure secure communications	Conduct surveillance in areas of potential insurgent activity	Provide accurate position location to U.S. advisors	Protect allied space capabilities		

Table 2.2

Illustration of Space Support for Anti-Terrorism, Crisis, and Country Conflict Operations

Type of Operation and Operational Objectives	ITW&AA	Weather/Environmental Monitoring	Space Launch	Satcom	Surveillance/Reconnaissance	Navigation/Positioning	Space Control	BMD	Force Application
Anti-Terrorism									
Locate and destroy centers of terrorist activity	Provide warning of potential missile launches			Ensure reliable, secure communications	Locate areas of terrorist activity		Ensure survivability of U.S. space assets		Destroy terrorist command and control centers through timely application of force from space
Crisis									
Monitor, assess, and respond to theater crisis	Monitor threat regions for missile launches	Assess weather in theater for possible air operations	Maintain or enhance on-orbit assets through timely launch	Ensure reliable, secure, and timely communication links in theater	Conduct surveillance in threat areas in support of developing theater air campaign	Provide accurate position location to theater forces	Identify potential threats to space systems		
Deter aggressive actions by belligerents		Assess sea state effects on enemy naval activities			Conduct surveillance on troop movements in theater	Neutralize hostile artillery		Initiate preparatory BMD actions	
Country Conflict									
Monitor situation to protect U.S. interests			Launch additional space capabilities as required	Relay critical information on situation		Maintain position location of possible belligerents	Protect ground stations, other space assets in country		

Table 2.3
Illustration of Space Support for Regional Conflict and Sustained Nuclear Operations

Type of Operation and Operational Objectives	ITW&AA	Weather/Environmental Monitoring	Space Launch	Satcom	Surveillance/Reconnaissance	Navigation/Positioning	Space Control	BMD	Force Application
Regional Conflict (MRC Level)									
Achieve and maintain air supremacy	Provide warning of threats to air assets	Assess weather for all air operations		Relay effects of air supremacy campaign	Locate threats to U.S. and allied air operations				Suppress enemy air defenses
Ensure access to space			Provide alternative launch capabilities				Protect space launch sites	Detect, track, destroy threats to space access	
Halt or evict invading armies				Ensure adequate and secure communications to respond	Determine routes of attack	Maintain position location of U.S. and allied forces			Deny enemy advances on battlefield by destroying command and control sites
Achieve and maintain sea control	Detect and track sea-launched missiles	Assess weather effects on sea control	Maintain on-orbit capabilities to ensure sea control	Provide secure, reliable communications	Maintain situational awareness of enemy naval forces	Maintain accurate position of U.S. naval forces	Protect space assets providing support to naval forces	Deter missile threats to naval operations	
Suppress war-supporting industry		Conduct adaptive route planning			Identify possible targets of interest				
Sustained Nuclear									
Provide unambiguous warning of attack	Detect, track, and assess missile launches			Maintain secure, reliable communication links to NCA and forces		Provide accurate position location f or responding forces		Initiate BMD actions	
Disrupt forces and war-supporting industry			Maintain survivable quick launch capability				Maintain survivability of U.S. and allied space assets		
Limit U.S. damage	Identify impact points	Assess effects of attack on U.S., allies		Reconfigure U.S. C3	Determine effects of attack on U.S. and allies			Defend against ballistic missile attack	Deny/destroy enemy means of long-range attack

For example, in Table 2.1, which illustrates peacetime, peacekeeping/humanitarian, and counterinsurgency operations, we see that in performing peacekeeping or humanitarian operations, an operational objective might be to establish and defend safe areas to ensure the safe passage of people as well as food and medicines. Operational tasks that might support the accomplishment of this objective using space-based systems might include the monitoring of theater ballistic missile (TBM) launches in areas of crisis (ITW&AA); ensuring secure communications between U.S. and coalition forces (satcom); denying infiltration in regions of concern (surveillance/reconnaissance); establishing accurate boundaries of safe areas (navigation/positioning); ensuring freedom of movement to, from, and in space (space control); monitoring potential threats to safe areas (ballistic missile defense); and deterring threat of attack against U.S. forces overseas (force application).

In Table 2.2, which shows anti-terrorism, crisis, and country conflict operations, we see that in periods of crisis, for example, an operational objective might be to deter aggressive actions by belligerents. Operational tasks using space-based systems might include assessing sea state effects on enemy naval activities (weather/environmental monitoring), conducting surveillance on troop movements in theater (surveillance/reconnaissance), neutralizing hostile artillery (navigation/positioning), initiating preparatory BMD actions (BMD), and denying enemy advances on the battlefield by destroying command and control sites (force application).

Finally, in Table 2.3, which shows regional conflict (MRC level) and sustained nuclear operations, we see that at the regional conflict level (MRC), for example, halting or evicting invading armies would be an operational objective that could perhaps be supported by the following operational tasks: ensuring adequate and secure communications to respond (satcom), determining routes of attack (surveillance/reconnaissance), and maintaining the position location of U.S. and allied forces (navigation/positioning).

Again, to illustrate the proliferation of space forces in today's world, the Cold War use of space forces would be devoted only to supporting the sustained nuclear operation cell of the table.

THE PROLIFERATION OF ACTORS IN SPACE

In the past, the U.S. military overwhelmingly dominated spacepower. Now, however, many actors (as shown in Figure 2.3) are involved in areas of space programs and activities that affect U.S. national security. These actors can be roughly categorized into four sectors of U.S. space activities: military, intelligence, civil (including scientific activities), and commercial. In addition, an

18 Space: Emerging Options for National Power

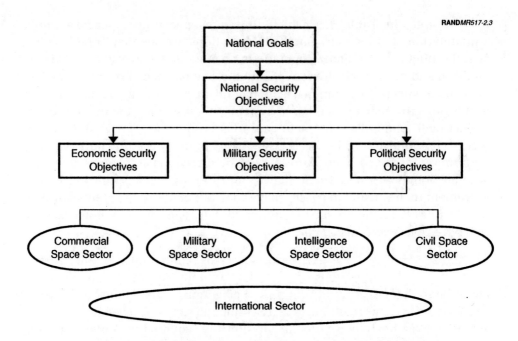

Figure 2.3—The Proliferation of Actors Involved in Space Activities

international sector has a strong bearing on the activities of the other four sectors. This international sector includes new capabilities (e.g., communication satellites (comsats), remote sensing) and management of those new capabilities by governments and nongovernmental organizations (e.g., other spacefaring nations, international organizations such as Intelsat, and regional organizations such as the European Space Agency) and their industrial counterparts (which in many instances are very closely linked with their governments). In the past it was sufficient to confine consideration of national-security-related space activities to the military and intelligence sectors, but this is no longer so.

Consideration of ongoing activities in each of the sectors and their influence on each other is necessary in light of budgetary and other factors. In fact, Figure 2.3 is somewhat misleading in that the sectors are much more closely linked than is indicated.

In addition, the interaction of these actors within the United States is exceedingly complex. Figure 2.4[11] illustrates the hierarchy and the complexity in the number of actors involved in making or influencing national space policy and

---

[11] Figure 2.4 and accompanying text were excerpted from unpublished work by Dana J. Johnson of RAND.

The "Proliferation" of Spacepower: A Geopolitical and Policy Context 19

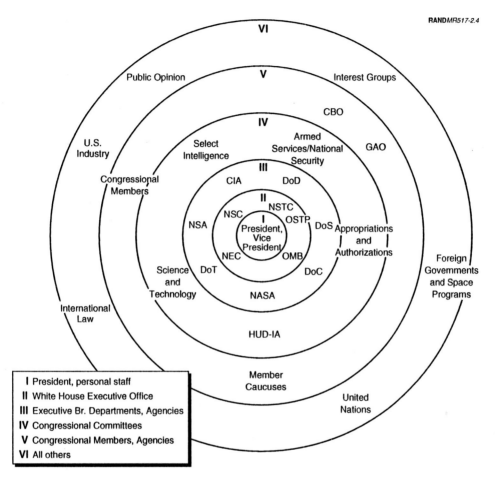

SOURCE: Dana J. Johnson, RAND, unpublished work.

Figure 2.4—A "Hierarchy of Levels" in the U.S. Government's Space Policy Network

military space activities. Six levels of decreasing influence or effect on the U.S. government's formulation of space policy are shown (with Level I, the President and his personal staff, being the greatest in terms of influence). Level II includes the various White House councils such as the National Security Council, the National Economic Council, and the National Science and Technology Council. Level III encompasses agencies having a programmatic self-interest in space policy, and Level IV focuses on Congressional committees having oversight or fiscal authority over space programs. Level V covers Congressional members and agencies, such as the Congressional Budget Office, the Office of

Technology Assessment,[12] and the General Accounting Office, which have addressed space-related issues at the direction of Congress. Level VI includes public opinion, interest groups, U.S. industry, the United Nations and international law, and foreign governments and space programs.

Delineation of these entities is not meant to imply that they act as single blocs; rather, the network is more ambiguous than is indicated. Different members within each group hold conflicting as well as similar opinions on certain issues, and thus cross boundary lines in the promotion of ideas and viewpoints. Although the scope of the network appears to be large, in reality there are probably only a few individuals who are truly influential in U.S. space policy.

---

[12] Disestablished in October 1995 by the Congress.

Chapter Three
TRENDS IN SPACE-RELATED FUNCTIONS: OPPORTUNITIES FOR COLLABORATION AND POSSIBILITIES FOR CONFLICT

Given the proliferation of space-related functions to support many military operations and actors, it is critical for the military to understand trends in these areas—e.g., who controls or leads in the area—especially as this information affects the development, acquisition, deployment, and exploitation of space systems providing information to the warfighter.

This chapter focuses on those trends and specifically examines areas that offer opportunities for mutually beneficial cooperation between military and civil/commercial sectors and areas that might raise concerns about conflict.

THE RELATIONSHIP BETWEEN SPACE-RELATED FUNCTIONS AND SECTORS

After examining the space-related functions shown across the top of Tables 2.1–2.3, it is clear that civil, commercial, and military interests have differing degrees of importance in each. At one end of the spectrum, as shown in Figure 3.1, satellite communications are largely driven by commercial interests in terms of numbers of customers, money, and, increasingly, deploying new technologies. At the other end, force application and ballistic missile defense are driven by military requirements, although they might use commercial technologies. In the middle are civil government functions related to public safety, such as monitoring weather. These positions are not static but can change over time as in the case of satellite navigation. For example, GPS was developed to meet military requirements but was found to have useful civil and commercial applications. In addition, military reliance on civil and commercial space systems implies a need to protect those assets in the Continental United States (CONUS) and in space. Space launch capabilities are considered to underlie all space activities and are, thus, primary concerns for all sectors.

Below, we discuss the status of each functional area in more detail.

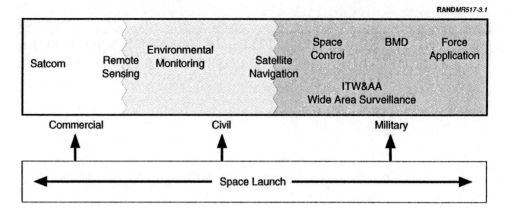

Note: Space systems only.

Figure 3.1—Space-Related Functions on the Spectrum Are Driven by Different Sectors

SPACE LAUNCH: GOVERNMENT-DRIVEN (IN TRANSITION)

Space launch is the necessary prerequisite for all aspects of spacepower. Intercontinental ballistic missiles for delivering nuclear warheads were pressed into service to become the nation's first space launchers. Over thirty years later, the technical descendants of those early launchers are still in service today, with the familiar names of Delta, Atlas, and Titan. Other vehicles were developed and abandoned along the way, most notably the Saturn family used for the Apollo program. The Space Shuttle was developed in the 1970s and will likely continue to be used for manned access to space in the early twenty-first century.

Through the 1970s, the United States was able to monopolize all free world launches. Europe made several efforts to create its own launch vehicle, but the technical and political problems of intergovernmental cooperation prevented success. In the mid-1970s, European disputes with the United States over launching European communication satellites (e.g., the French-German Symphonie project) gave a renewed impetus to the development of an autonomous space launcher. The result was the Ariane family of vehicles, developed by the European Space Agency (ESA) and marketed by the French-led consortium of Arianespace. When the United States experienced the losses of the Space Shuttle *Challenger* and several unmanned launch vehicles in 1986 and 1987, Ariane was able to take and hold a dominant share of the international launch market for several years (see Figure 3.2).

As the backlog of satellites resulting from the U.S. failures and stand-down declined and U.S. commercial launch providers began competing more successfully against Ariane, new competition arrived from China and, most

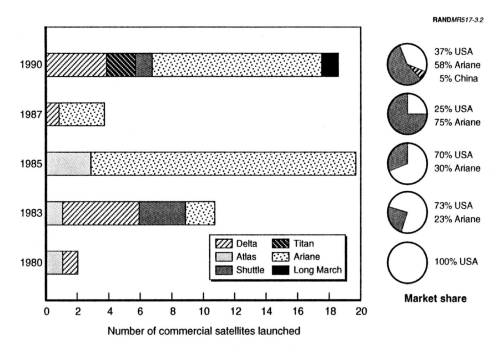

SOURCE: Reprinted from *Space Policy*, Vol. 9, No. 1, Bill C. Lai, "National Subsidies in the International Commercial Launch Market," pp. 17–34, copyright 1993, with permission from Elsevier Science Ltd., The Boulevard, Langford Lane, Kidlington OX5 1GB, UK.

Figure 3.2—The Space Launch Market

recently, Russia. As part of efforts to promote market reforms and to provide an incentive for other objectives, such as stemming missile proliferation, the United States entered into agreements allowing China and Russia access to the international satellite launch market.[1] This was done over the objections of the U.S. launch industry and Europe but with the support of U.S. satellite manufacturers, who welcomed more competition in launch services.

Decisions about space launch tend to last for decades and initiating dramatic change has proven very difficult. Even the former Soviet Union, which created more types of space launchers than any other country, practices a high level of technical inheritance and common design with new vehicles. There have been numerous reports and commissions over the years bemoaning the state of U.S.

---

[1] "Agreement between the Government of the United States and the Government of the Russian Federation Regarding International Trade in Launch Services," September 2, 1993, and "Memorandum of Agreement between the Government of the United States and the Government of the People's Republic of China Regarding International Trade in Launch Services," January 26, 1989.

space launchers, citing their operating costs, reliability, and unresponsiveness as causes for concern on military, commercial, and even scientific grounds (e.g., in launching science missions during narrow windows of opportunity).[2] An alliance of aerospace firms examined the possibility of lowering launch costs to stimulate new market demand and allow for the recovery of investments in lower-cost launch systems. The final report of the Commercial Space Transportation Study found that demand would indeed increase, but that dramatic and technically risky cost reductions (e.g., to $600 per pound for low earth orbit—LEO) would be required to dramatically increase the demand for space transportation and that financing could not be done solely by the private sector.[3] Space launch is often caught in a dilemma: Although virtually all observers find it unsatisfactory, no one has the combination of incentive and resources to pay for major new developments. The United States does have access to space, but the current costs of both satellites and launchers are so high that few resources are available for investments in new technologies or lower operating cost procedures.

The Clinton administration released a new statement of U.S. space transportation policy in 1994.[4] This policy covers the major areas of difficulty in space transportation, sets objectives, and defines agency roles, but it does not in itself solve the most basic problem—how new developments are to be funded. The DoD is charged with improving existing expendable launch vehicles, and the National Aeronautics and Space Administration (NASA) is charged with developing reusable space transportation systems, such as single-stage-to-orbit designs. The policy also calls for greater private-sector involvement in developing and operating space launchers, with the apparent hope that private-sector funding can be attracted to supplement government efforts. In part, the United States is looking to some sort of public-private partnership, whereby private industry would help pay for launch improvements in return for some combination of government co-funding and assured purchases to make an attractive commercial return. Such an arrangement would recognize that space launch improvements are in the interest of the civil, commercial, and military space sectors, but it is unclear whether a working partnership can really be created, given both financial and bureaucratic barriers for all parties.

The economic stakes in creating a successful partnership are high for both the DoD and private industry. According to Euroconsult's Launch Market Survey, the world space launch services market is expected to total more than $34 bil-

---

[2]*Report of the Advisory Committee on the Future of the U.S. Space Program* (1990); Vice President's Space Policy Advisory Board (1992).

[3]Boeing, Martin-Marietta, General Dynamics, Rockwell International, and Lockheed (1994).

[4]Executive Office of the President (1994a).

lion during 1997–2007.[5] In comparison, revenues were $18.3 billion during the previous ten-year period of 1987–1996. The driving force for this increase, beginning in the latter half of the 1990s, is the sharp increase in commercial demand for transportation to LEO. Major new satellite communications systems are being deployed in three market categories:

- "Big LEO" systems in the 1–2 GHz range, which provide voice and data communications, especially mobile telephone service (e.g., Globalstar, Iridium).
- "Little LEO" systems, which operate below 1 GHz and provide data communications such as e-mail, two-way paging, and messaging to remote locations (Orbcomm, Starsys).
- "Broadband LEO" systems, which provide high-speed data services such as videoconferencing and high-end Internet access primarily using the Ka-Band (e.g., Teledesic, Skybridge, Celestri).

According to the Department of Transportation's Office for Commercial Space Transportation, a "modest growth" scenario has 512 commercial LEO payloads being launched in the ten-year period from 1997–2006.[6] In rough numbers, this is equal to about half the number of all payloads launched worldwide during the 1970s or the 1980s.

The growth of commercial space demand means that the DoD can benefit from private capabilities but must recognize that it is an increasingly small part of overall launch activity. This continues a trend that has been under way for decades (see Figure 3.3). In the beginning, the DoD constituted a large percentage of U.S. and even worldwide space launches. This percentage declined as Soviet space launches built up and NASA activity increased. The end of the Soviet Union led to a rapid decline in the number of worldwide launches, thus increasing the relative percentage of DoD launches in the early 1990s. For the United States, about 80 percent of all launches in 1997 were commercial.

The growth of commercial space launches means that DoD will be an even smaller part of the world launch market in the future, despite the end of high levels of Soviet space launches. In 1997, the Aerospace Corporation conducted a study of future spacelift requirements for the United States through 2010.[7]

---

[5]"Space Launch Industry Faces Dramatic Changes" (1996), p. 86.
[6]U.S. Department of Transportation (1997), p. 6.
[7]The Aerospace Corporation (1997).

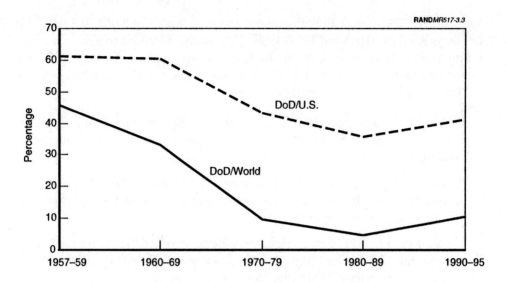

Figure 3.3—Percentage of Launches That Are DoD

Nominal estimates of annual launches for 2000–2010 were:

DoD launches	10–16
Civil launches	12–20
Commercial launches	42–87
Total	64–123

The percentage of annual U.S. launches for DoD would vary between 8–25 percent. This is quite a bit lower than for any earlier period of U.S. space activity. DoD is not alone, however; it is also estimated that the Ariane market share will decline to about 21 percent during 1997–2006 from about 50 percent in 1996.[8] This is about equal to the combined market shares expected to be held by the Chinese Long March and Russian Proton. In contrast to the fears attending the initial Russian entry into the market, U.S. firms such as Boeing and Lockheed Martin have created joint ventures to market the Zenit-3SL and Proton, respectively. The Sea Launch venture, which uses the Zenit, is expected to gain an 11 percent market share—equal to that of the Long March during 1997–2006.

The space launch market has been government-driven since it began. It became an international market in the 1980s with the introduction of the Ariane, and it will become increasingly international and commercial in the decade ahead. The demand for commercial rates of return, however, will make

---

[8]"Space Launch Industry Faces Dramatic Changes" (1996), p. 86.

it difficult for the private sector to finance completely new, reusable vehicles on its own. This means that the government, NASA, and DoD, should have an important role in developing new space launch technologies to lower the cost of access to space. On the other hand, technologies developed only for government needs are unlikely to be adopted by an industry that is responding to a large commercial market. Since no sector— civil, military, or commercial— seems able to "go it alone," some form of innovative, market-driven, partnerships will be needed if future space transportation opportunities are to be exploited.

Economic issues are clearly involved in space launch if the U.S. government hopes to attract private capital to new ventures. Economic interests also play an important role in two other aspects of space transportation policy: (1) the disposition of surplus ballistic missiles, and (2) stemming the proliferation of missile technology. Some U.S. companies and universities have proposed that they be allowed to convert ballistic missiles, made surplus by the START Agreement, to small space launchers. Although the number of payloads involved is not large, such proposals sparked protests from U.S. launch companies who felt that the U.S. government would in effect be competing with them by supplying surplus equipment. In addition, there were concerns that using ballistic missiles to generate revenue would make them attractive for sale on the world market.[9] As a result, the Clinton space transportation policy places a number of restrictions on the release of surplus missiles so as to not harm the commercial space launch industry.

Missile technology proliferation is a serious problem for U.S. security, because the technology for ballistic missiles is largely identical to the technology for space launch vehicles.[10] Thus, as countries develop their own space launch systems, they can also acquire the technology for building long-range ballistic missiles. Some countries seek to develop space launchers to have autonomous access to space, as Europe and Japan have. Other countries, such as India, Israel, and Brazil, also seek military capabilities as part of their space launch development programs. With the high costs and difficulties associated with space launch, it is not surprising that countries would seek to offset development costs by competing for commercial payloads. Thus, to the extent that the United States and Europe are not commercially dominant in space launch,

---

[9]As an aside, it is a sad commentary that twenty-year-old surplus ballistic missiles would be seen as a competitive threat to current launchers, but that is the state of space launch technology.

[10]These systems are not *technically* identical, but from the policy perspective of controlling missile technology proliferation, the U.S. position is that they are. The Missile Technology Control Regime (MTCR) is not intended to inhibit peacetime activities; however, similarities of propulsion and guidance, navigation, and control (GN&C) technologies and systems between intercontinental ballistic missiles (ICBMs) and space launch vehicles are a concern.

there are incentives for new countries to compete for launches. If these countries are not seeking a commercial rate of return, but revenues to offset the cost of developing a military capability, then they can offer very attractive prices. On the other hand, if the risks of competing commercially are seen as very high, domestic support for developing space launchers can wither, as happened in the case of South African efforts to market its own converted military launcher.[11] As an incentive for abiding by the terms of the Missile Technology Control Regime, the United States can offer to provide commercial launch services, as it has done for South Africa and Brazil.

If the United States could create a reliable, responsive, lower-cost space launcher, it would have not only better access to space but additional military and diplomatic leverage as well. It could lower the perceived value of surplus ballistic missiles on the world market, deter the development of competing commercial launchers, and offer incentives to stem missile proliferation. It could also strengthen the competitive position of U.S. satellite manufacturers by being able to offer "package deals" of launchers and satellites and to compete against the package deals of others (e.g., Russian offers to sell transponder capacity already on-orbit). Perhaps most important, it would allow limited DoD budgets to be spent more effectively on military space capabilities with operational benefits to the warfighter.

SATELLITE COMMUNICATIONS: COMMERCIALLY DRIVEN

Communication satellites (comsats) were the first commercial space success. Because of sales of satellites, ground stations, and transponder time, as well as the purchase of launches, comsats constitute the largest single source of revenue in the commercial space sector. Communication satellite technology was developed through DoD and industry funding, with NASA sponsoring the first demonstrations.[12] The United States currently has a lead in the systems integration skills required to produce communication satellites cost-effectively, but it is losing its lead in major subsystems and components to Europe and Japan.[13] Russian satellites, although technically inferior, are emerging as low-cost competitors on the world market for transponder time, through international companies such as Rimsat.

The DoD operates several communication satellite and "hitchhiker" payloads providing communication services. Examples include the Defense Satellite Communications System (DSCS), Air Force Satellite Communications System

---

[11]Sokolski (1993); Pace (1992).

[12]Cunniffe (1990); Hudson (1990).

[13]Edelson and Pelton (1993); Berner, Lanphier & Associates (1992).

(AFSATCOM), Leasat, UHF Follow-On (UFO), and Military Strategic and Tactical Relay (MILSTAR). These satellites are all in geostationary or high earth orbits, as are almost all commercial comsats. New proposals for commercial communication systems, such as Iridium and Globalstar, involve the use of large numbers of satellites in low earth orbits. The technology for new satellite communications, especially high-speed mobile services, is evolving so rapidly that the DoD is planning to make greater use of commercial systems rather than fielding its own systems. Advanced concepts for global battlefield communications, such as the Advanced Research Projects Agency (ARPA) Global Grid, share many of the same technical concepts and principles as civilian proposals to develop a national or global information infrastructure.[14]

The Defense Information Systems Agency (DISA) is the lead organization responsible for acquiring and providing comsat time. There continues to be debate within DoD and the services about whether DISA can or should continue to be a "one-stop shop" for providing non-DoD communications. Some argue that DISA is slow and unresponsive to demands for new communication services, especially in high-speed mobile applications, and that it is not flexible enough for the fast-moving, deregulated telecommunication market. Others argue that no other organization has the experience or defense-wide perspective to efficiently acquire the commercial communication services the military needs, whether from satellites, fiber, or cellular modes. Finally, there is also a question of whether DISA as a defense agency has the operational responsiveness for unified command requirements.

One of the most important uncertainties lies in the definition of realistic military communication requirements and joint coordination of those requirements. The technology is often evolving more rapidly than the current planning process and certainly faster than the acquisition process. Thus, a key question becomes: At what level should decisions about communication services be placed—when setting standards, buying connect time, or purchasing equipment? Authority could go to the services, could be placed at the U.S. Space Command (USSPACECOM) along with other space-related services, or could continue within the civilian offices of C^3I in the Office of the Secretary of Defense (with or without DISA in its current form). DISA has undertaken a Commercial Satellite Communications Initiative (CSCI), which will consolidate all DoD commercial satellite communication contracts, representing over $200 million in comsat services in 1994 alone.[15]

---

[14]Bonometti (1993).
[15]"DOD to Boost Civilian Satellite Use" (1994), p. 18.

The technical performance and cost benefits of using commercial communication systems seem clear, but there are risks as well.[16] Commercial systems can be more vulnerable to jamming, especially in the uplink, and their ground systems may be more vulnerable to attack than military systems.

The increasing commercial use of shaped beams using fixed K-band antenna technology aboard the satellite allows beam patterns that closely follow the contours of the service area. This means that power is not wasted on "unproductive" areas such as oceans and deserts. Unfortunately, U.S. forces do not always deploy to areas where commercial satellite services are concentrated. Even in areas where there is service, DoD may find itself competing for communications capacity with international media in which "spot prices" can become quite high. DoD may find itself considering innovations other than simple "rent or buy" decisions, such as taking equity positions in commercial systems with some priority rights that can be exercised or sold for capacity as communications requirements and technology evolve.

The spread of commercial satellite systems and international dependency on them can also create barriers to denying communications as part of economic sanctions in crises short of war. For example, during the Iranian hostage crisis in 1979, it was apparent that Iranian international telecommunications depended on access to Intelsat satellites. Intelsat was approached and asked to suspend service to Iran. Intelsat refused on the grounds that Iran was a member in good standing and there was no basis for denying it service. When pressed further, it was pointed out that if an Intelsat member could be ejected as a result of being unpopular or because another nation demanded it, then there may be enough votes to deny service to Israel as well. The United States withdrew its request and had to look for other pressure points to secure the release of the hostages.[17] Thus, although international organizations can be useful vehicles for cooperation and economic development, they can also be ill-suited to serve U.S. interests in times of crisis.

Similarly, the ability of the United States to ensure priority for its needs can also be undermined by the international participation required for communication systems in terms of both capital and market access. Thus, DoD use of commercial systems can bring technical and cost benefits but at the risk of additional vulnerability and inflexibility compared to completely government-owned systems.

---

[16] Bedrosian and Huth (1994).

[17] Broad (1980).

REMOTE SENSING: BECOMING COMMERCIAL

Space-based remote sensing for both surveillance and reconnaissance activities had its U.S. beginnings with aircraft, balloons, and even interplanetary probes.[18] The important scientific, military, and intelligence contributions of remote sensing are too numerous (and sometimes too classified) to treat here. Suffice it to say, the creation and maintenance of remote-sensing capabilities have taken many years, thousands of skilled personnel, and billions of dollars. Today, the most important factors driving space-based remote sensing are the simultaneous decline in support for traditional Cold War missions and the growing prospects for commercialization. The passage of the Land Remote Sensing Policy Act of 1992 and the release of Presidential guidance on operating licenses for private remote-sensing systems on March 10, 1994, have encouraged several U.S. companies to enter the competition for remote sensing.[19] This is a competition to provide remote-sensing services, but not to export satellites and especially not to export the underlying technology that remains subject to the restrictive export controls applied to munitions.

The ill-fated effort to privatize the Landsat program in the 1980s, in lieu of killing it outright, caused many people to believe that the remote-sensing market as a whole could not be commercially profitable. This led, in turn, to calls for government supports for civil remote sensing as a public good.[20] As the DoD gained experience with the multispectral imagery supplied by Landsat, it found the information useful for creating detailed maps and supporting combat intelligence functions. Multispectral Landsat data were combined with higher resolution, panchromatic data from the French SPOT satellite to create specialized but unclassified products for mission planning and other uses. DoD desires for a continuing source of good-resolution, multispectral imagery led to support for a High Resolution Multispectral Imaging Sensor (HRMSI) to be placed on the upcoming Landsat 7 satellite. Budget cuts led the DoD to pull out of the Landsat program in 1993 and the HRMSI effort collapsed, leaving the DoD to look for foreign and commercial sources of this type of data.

In the late 1980s, technical advances by DoD and the Department of Energy (DoE) opened up the prospect of smaller, lighter, more selective, and less-expensive remote-sensing systems. At the same time, computer costs were dropping rapidly and the market for geographic information systems was growing rapidly. With the costs of entry dropping and new markets for specialized remote-sensing data appearing, the potential for truly commercial remote-

---

[18]Davies and Harris (1988).
[19]U.S. Congress (1992); Executive Office of the President (1994b).
[20]Mack (1990).

sensing systems brightened. This led the Department of Commerce (DoC), which was responsible for Landsat and other forms of private remote sensing, to support streamlining the licensing process; as a result, several firms received licenses and were able to raise private capital.[21] These ranged from small firms such as WorldView Imaging, which received the first operating license since Landsat, to medium-sized space firms such as Orbital Sciences, and large defense contractors with a long heritage of military remote sensing such as Lockheed.

The arrival of multiple commercial remote-sensing firms poses a number of opportunities and risks for the military space community, particularly since the timeliness and quality of the products being offered would have once limited them only to the U.S. and Soviet governments. Commercial remote sensing offers the U.S. military potential new sources of remote-sensing data without requiring it to pay for the development of the space system. In a period of minimal new starts, it offers the potential for maintaining some crucial technical skills in the commercial sector that could no longer be supported on government contracts. If U.S. commercial firms can gain and hold a dominant share of the global market for supplying remote-sensing data, economics can deter new entrants and the proliferation of remote-sensing capabilities.

Among the risks posed by commercial remote sensing is the possibility of revealing to the enemy the disposition and movement of U.S. forces in times of crisis and war. There is also the potential for aggravating regional conflicts if hostile parties use commercially supplied information to make war on each other. The Orbital Sciences Corporation found itself entangled in both regional and domestic politics in the fall of 1994 when it announced that a Saudi Arabian firm would be an equity partner in its remote-sensing venture. Members of the U.S. House and Senate were concerned that Israeli security could be harmed and letters were sent to the DoC asking that various conditions be imposed on the venture and any associated export licenses. Orbital Sciences later pledged that Israel would not be imaged and that this could be enforced by the firm's precise digital control of its satellite's camera system.

The current issue that the DoD must face is the process by which it might seek to limit the operations of commercial and civil systems over sensitive areas or at sensitive times. The law and regulation of commercial remote-sensing licenses provide that the Secretary of State and the Secretary of Defense may ask the Secretary of Commerce to condition the operations of the licensees when national security conditions dictate. This authority has yet to be translated into a routine process, but extensive industry and government discussions are

---

[21]U.S. Congress (1993, 1994a, 1994b).

ongoing. Industry is willing to accept some restrictions as long as those restrictions are not seen as capricious and unreasonable. To the extent that DoD can provide advance warning or define preset areas for exclusion, business finds it easier to deal with limitations. For its part, commanders need to be aware of the potential risks of being imaged (as they have been for years with Soviet satellites) and to provide a process for getting proposed limitations on industry to the appropriate authorities in the DoD who can make the formal request to the DoC. In practice, this will also make it desirable to have prearranged exercises to ensure that the process actually works as intended. Of course, if U.S. industry fails to prevail in the marketplace, it is unlikely that the DoD would have any comparable say over the activities of Russian, French, or Japanese remote-sensing systems.

ENVIRONMENTAL MONITORING: CIVIL-GOVERNMENT-DRIVEN

Commercial ventures are not the only source of potential difficulty for the DoD in remote sensing. The National Oceanic and Atmospheric Administration (NOAA) in the DoC operates weather satellites in polar and geosynchronous orbits and participates in space data exchange agreements with other countries. Data from these systems are used to supplement information from U.S. military polar-orbiting weather satellites in the Defense Meteorological Satellite Program (DMSP). In the future, NASA's Mission to Planet Earth environmental research program will have a number of scientific instruments observing the earth, some of which might be expected to produce data of military value.

Weather satellite information is crucial to mission planning for all the armed services, as well as vital to civilian public safety and scientific research around the world. Both the DoD and NOAA have maintained separate weather satellite programs, with different orbital paths and instruments, although sharing a common bus structure. In 1994, the Clinton administration directed that the DoD's DMSP and NOAA's Polar Orbiting Environmental Satellite (POES) programs be merged into an integrated program office.[22] The purpose of the merger was to achieve cost savings by reducing duplication of effort. Proposals to merge the programs were made in the past but faltered largely as a result of disagreements over how to handle different civil and military requirements, such as nodal crossing times and data encryption, the latter point being a concern to the international meteorological community, which supports the free exchange of data. Technical integration is progressing, but concerns exist as to whether cost savings will be realized after paying for the transition.

---

[22] Executive Office of the President (1994c).

Weather satellites transmit their data to local ground stations and the resulting footprint can cover a wide area. During the Gulf War, U.S. commanders were concerned that Iraq would be able to use data from allied weather satellites, and efforts were made to deny downlink transmission to Iraq.[23] Unfortunately, this could not be done without also effectively denying weather satellite data to allied forces in the Gulf region. Instead, known weather satellite reception stations (and an unauthorized Landsat station) were destroyed during the allied air campaign.

Weather satellites are another example of dual-use information technologies employed in space, and economics plays an important role even though the field is not a commercial one. The debates over merging the DoD and NOAA weather satellite programs in light of the experience of the Gulf War point to the question of how to ensure that data go only to authorized users and nowhere else. These questions shape the differing military and civil views of data policy and how specific proposals, such as the inclusion of encryption capabilities, are received. The international scientific community has long shared data on a no-exchange-of-funds basis, even data from very expensive space systems. The United States has long been a major "supplier" of data and has been in a position to insist on a free system of exchange. As other countries put up their own satellites, notably the European EUMETSAT weather satellite system, there have been efforts to require access fees to help offset the costs of these systems. As the United States becomes more of a data "importer," its ability to insist on free data has lessened. Obviously, if all nations began charging for scientific data, the international scientific community would be hurt. When military planners seek to include encryption capabilities on civil weather satellites as a prudent measure, some foreign users see this a means of enforcing payment of access fees.

Another area of civil/military disagreement in the data policy debate is the use of declassified data from military space systems in environmental research. As senator, and then vice president, Al Gore has supported efforts to declassify data from military space systems and release them to the scientific community. An interagency Environmental Task Force with U.S. scientists was formed during the Bush administration to determine what kinds of data could and should be released.[24] As a result of experiments and advice from leading environmental scientists, many sets of historical intelligence data have been determined to have historical value, but policy questions on using such data for non-intelligence purposes remain. The commercial remote-sensing industry has

---

[23]Personal communication to S. Pace. The resolution of weather satellite images is measured in kilometers and Iraq would not have been able to detect the movement of ground forces.

[24]Studeman (1994), p. 24.

expressed concern over the release of declassified data, fearing a potential flood of government data that would depress the overall market.[25] U.S. industry representatives recommended that only data more than ten years old be released and only from systems no longer in operation. The latter point was made not only for security reasons, but also to ensure that the government would not be a continuing competitor in supplying certain kinds of data.

Whether the DoD continues to have its own weather satellites or instead relies on civil or even international systems, the U.S. government will have a significant influence on applicable data policy. Data policy issues such as fees for current or declassified data and the ability to selectively deny access to environmental data for economic or military reasons will be debated. DoD planners should recognize that there are multiple opportunities for both cooperation and conflict with civil and commercial users and will need to have an understanding of nonmilitary policy interests, such as the support of international science and the encouragement of a commercial remote-sensing industry.

SATELLITE NAVIGATION: COMMERCIALLY DRIVEN

The DoD operates a constellation of 24 satellites in 12-hour orbits that transmit precise time signals. Receivers in view of multiple satellites can use knowledge of the signals to calculate their positions and velocities anywhere in the world. These Navstar satellites make up the space segment of the Global Positioning System (GPS), which also consists of the ground control segment and the user equipment segment. The most precise signals, or P-Code, are reserved for U.S. military and other authorized users (e.g., allied militaries) by encryption key. A less-precise set of signals, the C/A code, can be used by anyone. The decision to allow civilian access was made by President Reagan in the aftermath of the Soviet downing of KAL 007 in 1983 and was intended to aid international air navigation.[26] The levels of accuracy available are described by the Joint Staff Master Navigation Plan and the biennial Federal Radionavigation Plan issued by the DoD and the Department of Transportation (DoT).[27]

There are U.S. and international proposals to augment GPS signals for aviation and maritime applications by installing additional beacons on the ground or in space to provide signal integrity monitoring and differential corrections for even greater accuracy than the P-Code signal (i.e., less than one meter). In fact, some commercial firms already provide encrypted differential corrections over FM broadcast bands and charge a fee for access to the encryption key in a

---

[25]U.S. House of Representatives (1994).

[26]Executive Office of the President (1983).

[27]U.S. Department of Defense (1994); U.S. Departments of Defense and Transportation (1992).

manner reminiscent of cable television. The Federal Aviation Administration (FAA), the U.S. Coast Guard, the U.S. Army Corps of Engineers, and the Department of the Interior (DoI) are all either creating or studying their own systems for providing differential GPS (DGPS) services, and the international mobile satellite organization, Inmarsat, has announced that it is interested in providing global DGPS services.[28]

The development of GPS cost more than $10 billion over two decades and reached initial operational capability in 1993. Even with the constellation only partially filled, GPS proved to be of immense value during the Gulf War. Virtually overnight, GPS receivers became a "must-have" piece of equipment for U.S. soldiers, sailors, and airmen. The demand for military GPS receivers exceeded U.S. production capabilities, but the DoD was able to use civilian receivers by making the P-Code available without restriction. The commercial GPS equipment industry began with land survey services in the mid-1980s when only a few GPS satellites were operating. The industry has grown, and is growing, rapidly—receiver sales alone were about a half-billion dollars in 1994 and are expected to exceed several billion in equipment and service sales by the year 2000.[29] Civilian and commercial sales are outstripping defense procurement of ground equipment and the user equipment industry is being commercially driven by fierce competition in electronics packaging, manufacturing, and software technology. GPS is being integrated with space-based communications and remote-sensing systems to create new civil and military capabilities.

GPS is increasingly important to the effectiveness of U.S. forces, and it is mandated for incorporation in all major weapons platforms by the year 2000, or those platforms will not be authorized.[30] Military commanders are increasingly sensitive to the health and state of GPS. When there are anomalies in GPS performance, the Master Control Station at Falcon AFB quickly hears about them from military (and civil) users worldwide. Although GPS was developed as a military system for military purposes, GPS has evolved into a dual-use information technology that is benefiting users in diverse fields, from surveying and aviation to vehicle tracking and mobile telecommunications. It may become a crucial utility for U.S. transportation infrastructure, global air traffic control networks, and advanced, high-speed communications, especially in mobile applications when position location is needed.

---

[28]Lundberg (1994).

[29]Interview with Michael Swiek, Executive Secretary, U.S. GPS Industry Council, Washington, D.C., May 1994.

[30]U.S. Congress (1993). GPS is required for inclusion in all DoD platform acquisitions by the year 2000.

A high degree of responsibility is being placed on the GPS program office and, by extension, on the DoD, to manage GPS well for military, civil, and commercial reasons. In narrow terms, commercial and military interests in GPS center around ensuring that the signal is neither misused or denied. Since GPS signal broadcasts from space cover very wide areas on the ground, it is impractical with current satellite designs to deny service to narrow regions of the globe. Thus, the denial of GPS signals to an enemy requires a secure encryption scheme for the P-Code and an appropriate jamming/spoofing capability for the C/A code. As a countermeasure to being denied, U.S. forces need a robust anti-jam, anti-spoof capability. Similarly, civil infrastructure potentially dependent on GPS, such as air traffic control systems, needs to pay attention to operational security against hostile and accidental threats.

In a larger perspective, a predictable and stable policy environment for GPS is necessary for it to become a global standard, to deter the proliferation of competing systems, and to allow U.S. industry the best chance of maintaining its current leadership position in growing commercial markets. GPS is finding rapid acceptance around the world, with Japan being the second largest manufacturer of GPS systems after the United States. The only comparable foreign system is the Russian GLONASS, but political instability and poor satellite reliability have hindered its international acceptance.[31] The United States could, however, incite the creation of competing systems if it fails to provide a stable, reliable GPS signal, if it attempts to manage GPS for the benefit of a specific industry, or if it attempts to charge for the basic GPS signals (which would be difficult and expensive to do).[32] If the United States continues to provide a free, high-quality signal, it is doubtful that anyone else will undertake the expense of building a comparable global space system.

U.S. control of the space segment of GPS allows for protection of U.S. military interests, while commercial competition in ground-based GPS equipment helps promote global economic growth. DoD can help promote U.S. economic and military interests by rapidly incorporating GPS into its own force structure and speeding foreign military sales to see that GPS is adopted by allied forces; by increasing awareness of operational security and electronic warfare threats within the services, among civil agencies such as the DoT, and among regional developers of GPS augmentations in Europe and Japan; and, most important, by continuing to support stable GPS operations as a high priority. Aside from

---

[31]N. Johnson (1994), p. 51.

[32]Nonetheless, indirect user fees could be assessed on specific industries benefiting from GPS augmentations, such as differential corrections supplied from ground stations to aircraft, ships, or cars.

missile warning, GPS is likely to be the most important ongoing military space program in terms of the effect of any system failure on the United States and the world.

Although GPS provides many economic benefits to the world, the spectrum band it uses is coveted by potential providers of mobile satellite services. At the 1997 World Radiocommunications Conference (WRC), which makes international spectrum allocation decisions, a coalition of European and Asian spectrum authorities advanced a proposal that would have allowed mobile satellite services to operate in a portion of the GPS band. This proposal was opposed by the United States as an interference threat to GPS signals and a threat to the use of GPS for safety-of-life services (e.g., international aviation and maritime navigation). The proposal was deferred for study only after a major U.S. diplomatic effort involving the highest civilian and military levels. GPS came under attack, not in a military sense, but as part of the intense international competition for spectrum in which billions of dollars are routinely at stake. The United States cannot assume that its current spectrum allocations will always be available to it, since it does not have a veto over international spectrum decisions such as those made at WRC-97.

INTEGRATED TACTICAL WARNING AND ATTACK ASSESSMENT: GOVERNMENT-DRIVEN

Throughout the Cold War, space-based and ground-based sensors kept watch for signs of attacking long-range aircraft and missiles heading toward North America. Space systems were tasked with detecting ballistic missile launches and nuclear explosions and reporting their locations to assess the scope of enemy and allied actions. This mission continues today in programs such as the Defense Support Program (DSP) and its infrared warning detectors and the nuclear detonation detectors housed on GPS. The performance of ITW&AA is vital to U.S. nuclear deterrence, and facilities such as Cheyenne Mountain are symbols of U.S. nuclear and space capabilities.

The missile warning and attack assessment mission continues to be important today, even with the decline of (if not elimination of) the likelihood of a massive nuclear exchange. The proliferation of weapons of mass destruction (WMD) and missile technologies means that a missile launch anywhere in the world has the potential of carrying such weapons, and the United States must be concerned about how such an event will affect its regional interests and the safety of its forces, citizens, and allies. The detection of theater missile attacks presents more difficult technical requirements than the launch of heavy ICBMs from the Soviet Union or China. First, the launches can come from a more diverse range of sites, both fixed and with the use of mobile platforms on trucks or trains. Second, the flight times of theater missiles are shorter and a greater portion of their flight is spent lower in the obscuring atmosphere. Third, the boosters themselves are not as "hot," i.e., their infrared signature is not as bright as that of ICBMs. These factors place a premium on the rapid detection of a launch and determination of the flight path. In some areas, such as the Korean peninsula, the possible range of flight azimuths is limited by the local geography, and, thus, ground and space assets can be effectively focused; however, other areas require dispersed and continual surveillance.[33]

As more nations face the potential threat of WMD-carrying missiles, the interest in their detection has also grown. Russia may face threats from its neighbors. Japan is concerned about potential tensions with China, Korea, and its more southern neighbors. Regional rivals, such as Iran and Iraq and India and Pakistan, are also concerned with potential missile attacks. In some areas, notably South Africa and South America, diplomatic efforts have been successful in lowering the threats of missile proliferation and WMD. Since it would be expensive and difficult for all nations to have their own global complex of space-based warning systems, there has been increasing interest in closer cooperation with the United States on missile warning. Such cooperation could involve the support of regional ground stations, the sharing of space- and ground-based data, officer exchanges, joint exercises, and even coproduction of hardware. Of course, the United States is more likely to enter into deeper cooperation with its traditional friends and allies and with those who share U.S. regional objectives. U.S. space capabilities provide a diplomatic advantage that no other nation can yet match and, thus, can be both a stick and a carrot for advancing U.S. nonproliferation and counterproliferation interests.

Unlike many other space capabilities, ITW&AA does not have direct commercial analogies in technology or mission. Although some of the component technologies have commercial uses or were initially developed as commercial systems (e.g., SBIRS satellite bus and software are based on Lockheed Martin's commercial TT&C product), the overall system requirements for tracking missiles and locating nuclear detonations are unique and exacting. From a policy viewpoint, ITW&AA is a central military responsibility and not something that can or should be placed in the private sector. That said, economic interests can and do play minor roles in this field. For example, the national interest in maintaining the GPS also supports the existence of a unique series of satellite platforms for nuclear detonation detectors that are vital to U.S. monitoring of nonproliferation agreements. DoD-supported research and development

---

[33]However, the short missile flight time in the Korean theater directly affects military operations and response times more than other theaters.

(R&D) for advanced infrared detectors may find spin-off applications in U.S. commercial remote sensing, where commercial markets cannot yet support such research alone. Perhaps most important, space-based warning systems provide new options for exercising national power in a post-Cold War environment that do not involve direct commercial competition. U.S. leadership in ITW&AA can provide new common bonds to other countries that are threatened by proliferation, and thus can contribute to a more stable international environment for all U.S. interests, whether military, political, or economic.

SPACE CONTROL: GOVERNMENT-DRIVEN

Ensuring free access to and passage in space, as on the high seas, has been a consistent objective of U.S. national security policy since the first Sputnik launch and is reflected in international agreements to which the United States is a party.[34] In the early space competition between the United States and the Soviet Union, free passage in space was more than a question of public prestige, but an integral requirement to the use of space-based intelligence systems. When the first Sputnik was launched, the Soviet Union did not request any permission to overfly other countries, including the United States. When the United States itself overflew the Soviet Union with satellites, Soviet objections were muted (certainly in comparison to protests over U-2 flights) as a partial result of the precedent that had been established.[35]

Having the right of unimpeded passage in space is not the same thing as being able to enforce that right, and the ability to deny the use of space to an enemy is the essence of military space control. Space control has been considered both offensively and defensively, encompassing antisatellite, survivability, and surveillance capabilities. Space control is often thought of in nautical analogies, such as maintaining sea lanes of communication or bottling up ships in harbor and preventing them from reaching the open ocean. Exercising space control in both offensive and defensive roles can encompass a variety of tasks such as launching an antisatellite (ASAT) weapon, jamming the communications of a hostile spacecraft, bombing hostile launch facilities and support structures, or applying alternative satellite survivability measures to reduce potential mission or functional vulnerabilities.[36] Space-related ground targets

---

[34]"Treaty on Principles Governing the Activities of States in the Exploration and Use of Outer Space, including the Moon and Other Celestial Bodies" (also known as the "Outer Space Treaty"), 18 U.S.T. 2410, T.I.A.S. No. 6347, 610 U.N.T.S. 2051, opened for signature on January 27, 1967. Article I states that outer space "shall be free for exploration and use by all States."

[35]Killian (1977); Steinberg (1981).

[36]For example, an earlier statement of National Space Policy discussed survivability in the following manner:

are likely to continue to be the most vulnerable parts of space systems because of the ease of reaching them and the relative difficulty of getting to space. At present, no nation possesses an operational ASAT capability that poses a significant threat to U.S. national security space systems.[37] In addition, the global reach of U.S. air and naval power provides the means for targeting the ground facilities of other spacefaring nations in extreme circumstances, thus achieving effective space control.

Having no immediate challengers to U.S. ability to access and use space cannot be counted on to continue indefinitely. As other nations acquire space launch capabilities and sophisticated guidance, navigation, and control (GN&C) systems or nuclear weapons, they may be able to pose additional threats to components of U.S. (and global) infrastructures such as commercial low earth orbit comsats and GPS-dependent air traffic control networks. The United States would certainly respond to a direct attack on itself or its allies, and although the question of a nuclear first use is beyond the scope of this work, the possibility remains that nonmilitary space systems could be targets.[38] As the leading spacepower, does the United States have an obligation or an interest in ensuring international access to space as it does in maintaining open sea lanes? How might the United States deter or respond to attacks on space-dependent communication and transportation networks? To stretch the analogy further, would the United States be willing to assert rights of free passage in the face of hostile claims, as it did in the case of Libya's claims over the Gulf of Sidra?

Like ITW&AA, space control is a military function and responsibility.[39] To the extent that commercial forces can help drive down the cost of access to space, the DoD will be able to exercise a more diverse range of space control operations (e.g., on-orbit inspections). To the extent that space commerce expands, the United States will have a growing interest in deterring hostilities in space. Space control is a means to an end, ensuring free passage in and through space

---

> DoD space programs will pursue a survivability enhancement program with long-term planning for future requirements. The DoD must provide for the survivability of selected, critical national security space assets (including associated terrestrial components) to a degree commensurate with the value and utility of the support they provide to national-level decision functions, and military operational forces across the spectrum of conflict . . . .

See the White House (1989), p. 11.

[37] Many nations have ground-based lasers capable of putting directed energy into space, and it is unclear what damage this could do to satellites in low earth orbit. Current lasers may therefore enable the creation of an improvised ASAT capability.

[38] The effect of the use of a nuclear weapon in regional conflicts has been treated in many different scenarios. See Millot, Mollander, and Wilson (1993).

[39] Commercial space suppliers have an interest in keeping their systems secure, but those interests do not necessarily extend to actively denying similar capabilities to their competitors (the "deny hostile use of space" aspect of space control).

in peacetime and denying hostile uses of space by an enemy in wartime. The threat of proliferating missile technology and WMD, and continuing regional tensions, mean that the United States has new opportunities for exercising post-Cold War leadership in support of the stable environment needed for global economic growth and as a counterbalance to aggressive regional powers.

FORCE APPLICATION FROM SPACE/BALLISTIC MISSILE DEFENSE: GOVERNMENT-DRIVEN

It has been argued that defending against ballistic missiles that pass through space is part of space control. This makes a certain amount of intuitive sense, since intercepting and destroying weapons passing through an environmental medium—be it air, sea, or space—is part of exercising control over that medium.[40] In practice, however, BMD tends to be treated separately from space control and further distinctions are made with respect to ground-based theater missile defenses (TMD) and space-based systems providing global protection against limited strikes (GPALS). The U.S. and Soviet-focused orientation of the Strategic Defense Initiative (SDI) has evolved into protecting U.S. forces overseas against ballistic missile attacks. Although no U.S. serviceman on the ground or at sea has been lost to enemy air action since World War II, the United States has suffered casualties from theater ballistic missile attacks, as happened in the Gulf War. As missile technologies spread, the ability of the United States to project force will come to depend on the ability to counter enemy missiles as well as aircraft.

As part of the transition out of the Cold War, it has become clearer that BMD technology could contribute to U.S. military objectives at various levels of conflict. In the near term, there is the need to protect deployed U.S. forces against attacks by theater ballistic missiles and to complement defenses against aircraft and cruise missiles. Depending on the particular situation, BMD systems may be asked to protect U.S. and allied forces, enhance regional crisis stability by offering protection from hostile, missile-armed states, or protect U.S. nuclear forces at home. These consist of a very wide range of missions, and the technology, not to mention the force structure, to perform them will be challenging to develop. Given the possibility that even a single missile carrying WMD could be devastating, U.S. leaders may eventually conclude that U.S. security cannot accept the risk of launch and a failed interception. How far they would be willing to go in using force, especially preemptive force, to achieve counterproliferation goals is still a subject for debate. Nonetheless, these concerns highlight the need to think beyond the threat of relatively primitive Scuds with conven-

---

[40]Johnson (1987); Hays (1994).

tional warheads. It may be that stealthy unmanned air vehicles or cruise missiles will present a greater long-term threat than ballistic missiles.

Ballistic missile defense is not a dual-use technology, although there is always the possibility of spin-offs from defense technology developments. Economics does play a role, however, in the effectiveness of BMD in deterring the spread of ballistic missiles or other long-range precision strike systems. If major investments can be negated by space capabilities that no other nation can match, then spacepower may allow the United States to ensure, with its allies, a more stable, safer world. If BMD cannot be made cost-effective, it can still be a useful capability for more narrow military objectives, but it will not be as effective a barrier to proliferation and the use of ballistic missiles. These are not solely technology questions, of course, but involve decisions about U.S. force structures and political will.

Moving from the defense to the offense, a potential far-term military space capability is that of "force application." This is usually taken to mean the use of kinetic energy or beam-energy-based weapons for attacks against space, air, ocean, or ground targets. The global coverage of space systems and relatively short times of flight could be advantages for space-based weapons, but the technical challenges of deploying, operating, and defending them are formidable. It is unclear if there are targets, other than ballistic missiles or command and control centers, for which the benefits of attacks from space are compelling. The technical problems are not so much in the weapons themselves (calculations have been done for numerous types of kill mechanisms) as in aiming. If a target can be found, it probably can be killed. The problem is finding the target accurately and rapidly. This, in turn, begs the question of whether it might be more effective to use space-based systems to find targets and cue terrestrial "shooters" rather than using space-based weapons. Apart from technical issues, however, a more fundamental aspect to the issue of space-based weapons is the political will necessary to make such a step possible.

Force application missions appear to have the least relation to U.S. economic issues and interests. Given the decline of the Soviet threat, the strategic applications of space-based weapons appear to be limited.[41] Force application tasks might instead be used as "silver bullets" or as special forces in accomplishing unique, high-priority missions that are not possible with conventional military forces. Given the expense and technical demand involved, however, space-based weapons are not likely to be built for lower levels of conflict unless the

---

[41]Others, however, might argue that the end of the Soviet Union leaves the United States with greater freedom to employ space-based weapons.

cost of access to space declines dramatically and the sophistication of sensors and information processing grows in a similarly dramatic way. Such changes are certainly possible, but in the current budget, political, and threat environment, it is unlikely that the United States will make a decision to develop space-based weapons in the near future.

SUMMARY

Table 3.1 summarizes the space-related functions discussed above in terms of which sectors dominate, as well as the various opportunities and problems DoD will confront within those areas. Although DoD does dominate many functions (such as ITW&AA, space control, and force application), it is often a follower in other important functions (such as space launch, satcom, and environmental monitoring). This situation has important ramifications for the kinds of space force structure DoD pursues or might consider pursuing, and for the military's ability to support national security objectives. Consequently, it is important for DoD to consider alternative military space strategies for the post-Cold War world, which is the subject of the next chapter.

Table 3.1
Summary of Space-Related Functions and DoD Opportunities and Concerns

Space-Related Functions	Sector	Leading Players	Potential DoD Opportunities	Potential DoD Concerns
Space launch	Civil leads; small commercial	Europe, with U.S. following; other countries involved	If civil/commercial can produce reliable, responsive, lower-cost launcher, can increase access to space and improve military/diplomatic leverage	Requires that financial and bureaucratic barriers to civil/commercial collaboration be overcome
Satcom	Commercial leads	U.S., with Europe and Japan closing gap	Can leverage off growing commercial capabilities	Is vulnerable to jamming/attack and to international growth, which creates barriers to denying enemy use and ensuring U.S. priorities; DISA not responsive enough to be "one-stop-shop"
Remote sensing	Becoming commercial from DoD/civil origins	U.S., with other countries involved	Can leverage off growing commercial capabilities	Needs to limit operations over sensitive areas/times; information used by hostile parties can aggravate regional conflicts
Environmental monitoring	Civil	U.S., with other countries gaining (EUMESTAT)	Can leverage off civil capabilities	Needs to ensure that data go only to authorized users; needs to balance free access to declassified military data with market concerns of remote-sensing industry
Satellite navigation	DoD (space segment); commercial (user equipment)	U.S., with Japan and Russia following	Can control and maintain the function	Needs to balance military, civil, and commercial needs; needs to ensure that GPS is the global standard to avoid proliferation of competing systems
ITW&AA	DoD	U.S.	Can provide leadership to countries threatened by proliferation of WMD and missile technology, and contribute to ensuring global stability	Needs to balance critical security interests and interests of allies and partners, coalition operations
Space control	DoD	U.S.	Can provide leadership to countries threatened by proliferation of WMD and missile technology, and contribute to ensuring global stability	Needs to be aware of potential threat to U.S. space infrastructure as other countries gain space launch and GN&C systems, and growth of multinational commercial systems expands
Force application/BMD	DoD	U.S.	Can help ensure more stable world	Requires political will to deploy capabilities; potential treaty compliance issues

Chapter Four

ILLUSTRATIVE MILITARY SPACE STRATEGY OPTIONS IN THE POST-COLD WAR WORLD

As we have shown in Chapter Two, U.S. military space activities have undergone a radical transformation in the post-Cold War world. In the past, these activities were driven by strategic nuclear (e.g., missile warning) and intelligence functions. Even more important, these functions were traditionally directed at the same target, the Soviet Union, with attention to other regions as dictated by the need to contain local efforts of Communist influence. It is not much of an exaggeration to say that the requirements for space systems were as simple as "provide all available information on what is happening in this denied area" or "provide warning of an ICBM attack within X seconds of launch." Today, the requirements for space systems are far more complicated and entail the performance of multiple functions in support of multiple missions and operations. To be effective in this world, the military space community will need to move beyond its roots in the nuclear and intelligence communities and be supportive of new demands from conventional military forces.

At the same time, the U.S. military must recognize that it is no longer the sole actor in meeting national security needs and objectives. In fact, as shown in Chapter Three, for some of the functions space systems are likely to perform in supporting military operations in the post-Cold War world, the military is a follower rather than a leader.

The real question is how space forces will support U.S. military strategy within this context of multiple functions and multiple actors, both military and commercial, in this country and abroad. Ultimately, as a part of U.S. military strategy, U.S. military space strategy will need to support U.S. national space objectives—objectives that are increasingly being driven by the need to balance military and economic interests in a globally interconnected world.[1]

___

[1] The case of the U.S.-Japanese collaboration to develop the FS-X fighter illustrates the need to balance military and economic interests in the context of national security interests. From a military point of view, keeping the Japanese a functioning and integrated part of the U.S. military posture in

In this chapter, we examine a spectrum of illustrative space strategy options that might emerge. We then consider the implications of those space strategy options in terms of meeting the needs of national space objectives.

ILLUSTRATIVE MILITARY SPACE STRATEGY OPTIONS

In thinking about future military space strategy options that could emerge, we developed a spectrum of three broad options:

- *Minimalist:* The military use of spacepower is highly dependent on external relationships and partnerships. Integration with other military operations depends on organizations outside the military chain of command. This strategy option is largely the outcome of budgetary constraints and technological advances in other sectors, thus leading to the U.S. military owning only those systems that perform unique and/or critical national security functions and leasing everything else from the commercial sector.

- *Enhanced:* The military use of spacepower is highly integrated with other forms of military power. External relationships and partnerships are important but not critical to core military capabilities.

- *Aerospace Force:* Military spacepower is exercised through an independent service and is fully capable of being exercised separately from other military forces. Actual military operations are most likely joint and combined and may use external relationships, but this is not required.

These three options are intended to be evolutionary, rather than revolutionary, in nature, and are assumed to build from the status quo of forces and budgets. They are also not intended to be comprehensive strategies, for there could be many variations within each strategy as well as distinct organizational and functional differences among all three. For example, the Aerospace Force option could occur as the result of the Air Force evolving into a true aerospace service. Other options that exploit revolutionary technologies and capabilities are also conceivable and could be radically different in nature and outcome from the three evolutionary options discussed here. It is not clear that one option is necessarily better than the other, or that one is easier to manage and organize than another. For example, because of the necessity within the Minimalist strategy for the military to work closely with industry, that strategy

---

the Pacific argued for transferring appropriate U.S. aerospace technology to ensure that the fighter developed was compatible with U.S. military needs. However, economic concerns over losing vital aerospace technology and enabling Japan to compete with the United States in commercial airplane development argued for denying the transfer of aerospace technology to Japan. In this case, both sides were legitimately arguing that their position was based on the need to maintain national security objectives.

might be more able to fulfill U.S. economic objectives than the Aerospace Force strategy option.

Figure 4.1 illustrates the distinctions among the three options from an organizational perspective. The left side of the chart shows increasing levels of functional capability, vice increasing levels of DoD involvement (across). Those functions near the bottom (excluding satellite control/TT&C, which is common to all) are considered a high national security priority, and those at the top are partially linked to future changes in national objectives and priorities. The figure does not get into procedural and other matters such as budgetary and acquisition authority, or command and control relationships. Furthermore, it should not be assumed from this figure that agencies such as NASA go away in the Aerospace Force option; rather, such activities as science and planetary exploration are assumed to continue in some form (but they are not discussed in this document).

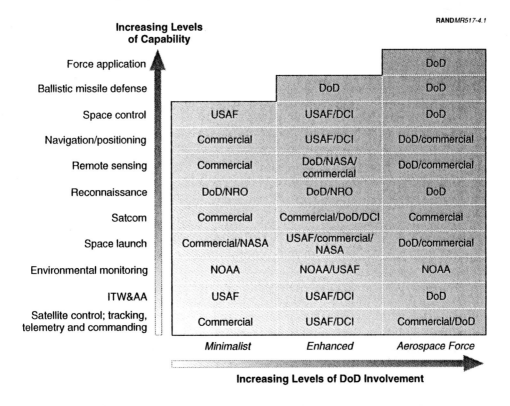

Figure 4.1—An Organizational Perspective on the Three Strategies

The reader will note that Figure 4.1 draws a distinction between ballistic missile defense and force application, whereas Chapter Three chose to combine them. The rationale beind this difference lies in perspective. The commercial world views the use of deadly force as properly an exclusive governmental function; however, from a national security perspective, BMD and force application are distinct functions with associated objectives and tasks requiring the use of military force.

The Minimalist Strategy Option

Assuming that the defense budget continues its present rate of decline, the DoD, and the Air Force in particular, will be faced with force structure triage decisions. Notwithstanding various visionary expressions of the future of the Air Force in space, the Air Force's first priorities would likely continue to be reserved for aircraft.[2] Taken as an extreme case, the Minimalist option results from moving as many space functions as possible from military to civil or commercial ownership and forgoing the development of new capabilities, such as ballistic missile defense from space. Space systems with dual-use roles, such as communications, remote sensing, and satellite navigation, are transferred to commercial or civil control. Weather satellites (and likely the Navstar global positioning satellites as well) are under civil government control, since, in this option, maintaining the space segment is not commercially viable without large government commitments. NASA and commercial space launch vehicles would provide DoD access to space as needed. DoD would retain the ITW&AA and reconnaissance functions as necessary for nuclear deterrence, theater operations, and as part of international commitments to North American defense under the North American Aerospace Defense Command (NORAD).[3] The Air Force would likely retain a core of technical expertise in space, but capabilities in the other services would probably wither and be limited to developing and fielding user equipment.

The Minimalist strategy is driven by increasingly severe defense budget constraints and potential technical advances that make it more cost-effective to use commercial rather than government systems for space-based information needs. In effect, as shown in the figure, the only functions directly under mili-

---

[2]Serious problems arise from focusing on the instruments of airpower, rather than the doctrine and rationale for airpower. In a similar manner, it would be a mistake to focus on space systems rather than on spacepower and missions served by those systems. Builder (1994), p. 281.

[3]The Minimalist option assumes that the reconnaissance function is maintained by the DoD/NRO. Conceivably, the activities conducted by the NRO could be combined organizationally into the new National Imagery and Mapping Agency or some other entity. However, we have chosen to include the NRO in this analysis.

tary control are ITW&AA and aspects of space control, primarily space surveillance; the remaining functions have either been split off from the military and comanaged with other government agencies (e.g., weather satellites with NOAA) or leased from the commercial sector. Also, the USAF (and the rest of DoD, for that matter) will rely exclusively instead on commercial and NASA launch capabilities.

In some cases, it might be argued that leasing communications capabilities is more expensive than owning. Although a primary and even necessary motivation for choosing a Minimalist option might be to reduce government expenditures, the strategic choice is one where the DoD depends heavily on others for spacepower. A more difficult choice may the case of GPS. All critical national infrastructures (e.g., transportation, telecommunications, oil and gas, and electrical) are increasingly dependent on GPS. Current Presidential policy requires GPS to remain responsive to the National Command Authority in part because of the importance of GPS to both civilian and military users. Thus, while it may be possible to imagine satellite navigation signals provided by a commercial or international entity from a service perspective, it is difficult to imagine this occurring from a national perspective. No other system acts like GPS in terms of being both a global and U.S. "embedded utility." Thus, GPS is assumed to remain under civil government control even in the Minimalist option.

The Enhanced Strategy Option

The strategic choice represented by the "Enhanced" option is to enhance spacepower by aggressively integrating spacepower wherever feasible into all types of military capabilities and operations.

The Enhanced strategy includes more aggressive efforts at reforming service roles, missions, and functions in space, developing joint space doctrine, and taking a more proactive approach to civil/military space cooperation. It also includes the intelligence sector as an integrated part of national security space activities. If there are no new space initiatives and the defense budget stabilizes, then the Enhanced strategy option represents a continuation of current practices, albeit with more integration between military and intelligence activities and functions, and with some increased capabilities to deal with rogue or Third World missile threats. Consistent with National Space Transportation Policy, the DoD would lead in developing improvements to expendable launch vehicles, while NASA focuses on reusable launch vehicles.[4] Civil, commercial, intelligence, and military space sectors would retain independent access to

---

[4]Executive Office of the President (1994a).

space as well as their own communications and remote-sensing systems. As a result of consolidations in polar-orbiting weather satellite programs, the civil government sector would gain a greater role in environmental monitoring. In satellite navigation, nonmilitary uses would grow faster than military uses, but GPS's importance to U.S. forces would keep it within the DoD while allowing wide civilian access. Building on work under way in the services, at USSPACECOM and its components and in Tactical Exploitation of National Capabilities (TENCAP), tactical applications of strategic nuclear and intelligence-oriented space systems would continue to be developed for conventional and special forces missions. The Air Force would continue as the primary developer and operator of space systems with smaller cores of expertise in the other services growing with time.

Current airborne capabilities could be moved into space. For example, AWACS and JSTARS functions might be performed by constellations of radar satellites.[5] This would require technology advances in space-based power and the precise control of large structures, but no scientific breakthroughs. The size of the constellation would depend on the operating altitude and coverage area required: More satellites would be needed at lower altitudes, but each would require less power. The level of effort would be comparable to deploying a mobile satellite communications system.[6] The benefit of moving to space would be avoiding the costs of maintaining aircraft and crews. One risk would be ensuring that the system is at least as survivable as current AWACS and JSTARS, if not more, given the expense of development and deployment.

New military capabilities would be needed to deal with problems created by reliance on space systems. In the case of GPS, work on NAVWAR capabilities is already under way to insure that U.S. forces can acquire and maintain access to GPS signals under conditions of jamming and spoofing and to deny unauthorized users access to GPS signals. Given increasing civilian and commercial reliance on GPS, care will have to be taken that NAVWAR effects are limited to the actual area of conflict. NAVWAR will also pose challenges for allied and coalition partners who need to be interoperable with U.S. forces. Thus, the challenges of NAVWAR will be organizational as well as technical.

Some existing commercial capabilities, such as satellite communications, will need enhancements that are of interest only to the military. These include

---

[5]The AWACS function would be to maintain surveillance of F-16 and larger-sized aircraft in six 350-nm circles with a ten-second revisit rate. The JSTARS function would be to maintain surveillance of slow-moving targets on two Army corps frontage areas, each about 200 km on a side. The revisit interval can be about a minute, but the radar must be able to discern lower velocities for targets such as trucks and helicopters.

[6]Hagemeier et al. (1995).

protected comsats that are jam-resistant, anti-scintillation, and capable of low-probability-of-detection and low-probability-of-interception communications. Whether these capabilities can be added to commercial systems or whether they would continue to require dedicated military systems would remain to be determined.

If the DoD is able to have a stable budget or slight real growth, it could make greater use of commercial and civil space assets to enhance its current military capabilities. The Enhanced strategy option could see a stronger role for the Commander-in-Chief, U.S. Space Command (CINCSPACE) as the DoD focal point for civil/military operational cooperation (as opposed to policy coordination), greater attention to common or joint doctrine for space operations, and coordinated DoD efforts to foster U.S. commercial space capabilities to strengthen U.S. national capabilities in space. In satellite communications, one might see extensive DoD use of new low earth orbit communication satellite systems and purchases of high-resolution remote sensing data. For access to space for unmanned payloads, DoD would use commercial launch vehicles and not NASA's manned spacecraft (i.e., the Space Shuttle). Modest budget growth in space systems could accommodate space-based portions of a ballistic missile defense (e.g., the space-based infrared system (SBIRS)) and new space control capabilities. However, modest growth would not likely be able to support the development of space-based weapons. From an institutional viewpoint, the Air Force and other services would continue their existing space-related activities but with more attention to joint requirements. DoD civilian leadership would likely play a larger role in coordinating DoD space activities with other national policy objectives, particularly in economic strategies.

Aerospace Force Strategy Option

At the other end of the spectrum from the Minimalist option, the Aerospace Force strategy consists of independent, robust military space capabilities in critical national security and warfighting functions, supplemented with mechanisms to ensure access to existing civil and commercial space capabilities, on which the DoD is already dependent, in times of crisis or war. In this option, the DoD would still rely on commercial space launch capabilities for the bulk of its routine peacetime requirements, but it would maintain quick-response capabilities (e.g., transatmospheric vehicles or similar technology) to ensure autonomous access to space. Other capabilities that would be either DoD-owned or DoD-dominated would include ITW&AA, reconnaissance, navigation, space surveillance, space control, and ballistic missile defense.

The Aerospace Force strategy would include the development of new capabilities such as space-based weapons for force application purposes, consistent with the rationale of identifying new vital interests in space and the protection of existing treaty and alliance commitments. Such efforts would obviously require dramatic increases in military space budgets and would also represent a possible transition in thinking about the use of *aerospace power* as a continuum to accomplish national objectives versus *air and spacepower*. In this instance, the Air Force could evolve into an Aerospace Force, focusing its institutional interests on the development and exploitation of aerospace power with limited air-related core competencies (e.g., airlift) and possibly transitioning other noncore capabilities (e.g., close air support) to the other services, or the Aerospace Force could become a new independent service. However, an independent service would likely run into institutional, budgetary, and bureaucratic roadblocks from the other services, and, as is true with the other options, care would have to be taken to ensure that the interests and needs of all services for space products and information, as well as a broad range of air support, are met.

In the remainder of this chapter, we will address how each of these strategies would accomplish national security goals and objectives, and how they would perform the missions described in Chapter Two; e.g., what does it mean to perform the peacekeeping mission with a Minimalist strategy in place? We will also examine several factors that will influence outcomes under each of the strategies, such as trust and space literacy.

HOW THE OPTIONS ACCOMPLISH SPACE-RELATED NATIONAL SECURITY OBJECTIVES

As enunciated in the National Security Strategy, U.S. national security objectives pertaining to space operations include the following:

- Ensuring continued freedom of, access to, and use of space;
- Maintaining the U.S. position as the major economic, political, military, and technological power in space;
- Deterring threats to U.S. interests in space and defeating aggression if deterrence fails;
- Preventing the spread of weapons of mass destruction to space;
- Enhancing global partnerships with other spacefaring nations across the spectrum of economic, political, and security issues.[7]

---

[7]See the White House, n.d., p. 10.

Any potential options such as the three described above should be capable of effectively supporting these objectives. However, *how* they support these objectives—how they carry them out—is of interest. We pointed out in Chapter Two that the sectors of space activities (civil, commercial, military, etc.) are much more intertwined and interdependent today than they were during the Cold War. Consequently, there is a greater necessity to understand the linkages among political, economic, and military strategies to accomplish national security objectives. The objectives of national defense, to fight and win the nation's wars, are obviously not those of economic policy. Yet economic strength is of vital importance to being able to provide the material resources of defense, and commercial R&D is of increasing importance to providing the technical superiority required by the American way of war.[8] Access to capital and technology, to information flowing across global communication networks, and to rapidly evolving markets for sophisticated goods and services are vital issues to U.S. business. What the military decides to buy, how it decides to buy it, what R&D it is willing to support, and what position it takes in the interagency development of U.S. policy affect both military and economic interests.

The national security community can affect U.S. industry by how it chooses to balance economic interests with traditional security interests when the two are competing. In cases where economic and traditional security interests are complementary, the question becomes whether it is more effective to promote the prospects of a U.S. firm or industry or create obstacles to competing foreign interests. It is therefore prudent to consider the accomplishment of national security objectives through a range of alternative means, some of them being not the application of force but the "shaping of the battlefield" ahead of time, which may result in the same desired outcome without the potential escalation, and penalties, of conflict. Traditionally, the term "shaping the battlefield" applies to the use of preattack actions before the initiation of a military offensive, including the use of artillery and air bombardment to weaken enemy defenses. In this context, the term is intended to consider the use of nonmilitary techniques to constrain or restrict what weapons and military options the adversary has before the onset of hostilities.

What follows is a discussion of how each option would accomplish the space-related national security objectives outlined above. The discussion is intended to be illustrative and conceptual, not evaluative. The United States could choose to pursue aspects of one or all of these strategies. A summary of the discussion is shown in Table 4.1.

---

[8]See Weigley (1973) for a discussion of the role of technology in the American military; and Keegan (1993) for a longer perspective on the role of changing technology in the culture of warfare from ancient to modern times.

Table 4.1
How the Options Support Space-Related National Security Objectives

National Security Objectives for Space	Minimalist Strategy Option	Enhanced Strategy Option	Aerospace Force Strategy Option
Preserving freedom of, access to, and use of space	Warning and reconnaissance only; high premium on survivability of critical space-based assets; enforcement through other means, e.g., economic strategies, terrestrial military forces, treaties	USAF maintains space launch capability, supplemented by NASA, commercial launch; space-based surveillance necessary; high premium on space control capabilities (including survivability measures)	Strong space control, force application capabilities with space launch enable protection of critical space assets and ensure access to commercial space capabilities
Maintaining U.S. economic, political, military, and technological position	Dependent on U.S. commercial developments; reliance on integrated use of other instruments of national power	Emphasis on multi-sector coordination and cooperation; coordinated inter-agency policy development and implementation	Military/civil/commercial cooperative approach necessary; interagency policy and strategy to provide guidance
Deterring/defeating threats to U.S. interests	Warning and reconnaissance only; high premium on survivability of critical space assets; use of other non-space forces required	Integrated sector approach necessary; space control, limited BMD to deter threats, backed by terrestrial forces	Dominance of aerospace medium through use of air and space-transiting vehicles and weapons
Preventing spread of WMD to space	Requires integrated policy approach using variety of means (economic, diplomatic, military)	Requires integrated policy approach backed by space control, limited BMD	BMD and force application capabilities could deter; also "shaping the battlefield" through use of economic incentives not to build or deploy space launch vehicles
Enhancing global partnerships with other spacefaring nations	Global partnerships to accomplish national security objectives are commercial in nature rather than military, but tempered by competitiveness issues	Proactive approach to U.S. civil/military space cooperation, and commercial space developments strengthen U.S. position in partnerships	Capabilities enable fulfillment of alliance and coalition relationships and protection of U.S. and allied space-based and terrestrial assets

Accomplishing National Security Objectives: The Minimalist Strategy Option

Preserving Freedom of, Access to, and Use of Space. As mentioned above, ITW&AA, reconnaissance, and some space control functions are the only elements of this option under DoD control (more specifically, Air Force control). Therefore, there is a high premium on ensuring the survivability and protection of those assets, perhaps through the use of terrestrial forces. Accomplishing this objective in this option would be through nontraditional means (i.e., economic and commercial strategies, treaties, and other mechanisms). Elements of alternative commercial strategies include (1) the use of voluntary contracting to ensure access to commercial products and services, much as CNN leased transponder time from INTELSAT during military operations in Somalia; (2) obtaining desired features from other, nonspace elements of commercial satellite systems by joining with other governmental and nongovernmental users that share common needs; and (3) making better use of existing satellite systems—both commercial and military—by improving DoD's policies, procedures, and priorities.[9] These elements imply dealing with the issue of trust, or lack of it, among space sectors by developing measures to demonstrate DoD's commitment to being a "good customer" and the ability of the Air Force, as the primary service provider of space-based functions, to meet other services' and users' needs.[10]

Maintaining the U.S. Position as the Major Economic, Political, Military, and Technological Power in Space. To the extent the U.S. military can use commercially developed hardware and technologies, it does not have to develop and maintain those parts of its defense industrial base, except to be a good customer. If some commercial technologies are superior to or more readily available than military counterparts, the effectiveness of U.S. forces can be improved (e.g., as has been the case in the use of personal computers). More subtle is the use of economic forces to shape the future environment in which U.S. forces operate. For example, if the United States could lead in providing commercial satellite communications, remote sensing, satellite-based navigation, and space launch services, it would be in a better position to benefit from those national capabilities in wartime. If the United States cannot peacefully influence how and when those capabilities are provided, it may have to escalate the use of military force to protect its interests. In terms of strategic nuclear conflict, economic power in dual-use capabilities provides additional escalation options to the National Command Authority without requiring additional force structure. Under the terms of this option, these considerations would likely be a necessity.

Deterring Threats to U.S. Interests in Space and Defeating Aggression If Deterrence Fails. If U.S. space systems are to dominate the commercial and military

---

[9]See Poehlmann (1996) for a discussion of the applicability of the "CRAF [Civil Reserve Air Fleet] model" to the issue of ensuring access to commercial space capabilities in times of crisis.
[10]See Johnson et al. (1995).

uses of space, then preventing the misuse or denial of these space systems in times of national emergency is critical. Dependence on a few systems for access to, and use of, space can be a source of great leverage or vulnerability depending on how reliable and secure those systems are. The ability to discriminate among users is obviously important to commercial systems so they can collect revenues. This same discrimination ability is important to military systems to ensure that allies get the support they need and that enemies are denied or deceived. The challenge for military planners in this option is that they cannot necessarily count on commanding non-DoD space systems to discriminate. Non-DoD space systems are continuing to increase in the U.S. commercial and civil sectors, as well as in other spacefaring nations. The technical means of discrimination, whether by jamming, spoofing, encryption, or other means, are often less difficult than ensuring that decisionmaking processes to employ such discrimination mechanisms exist that are mutually acceptable with U.S. civil and commercial space systems. Thus, if the United States seeks to have secure and reliable space systems, it will need to foster cooperative agreements between DoD and the private sector, as well as with civil space agencies such as NASA and NOAA. Similar cooperative agreements can also be considered for allied space capabilities.

Preventing the Spread of Weapons of Mass Destruction to Space. Deterring the spread of ballistic missile technologies is beneficial to both U.S. economic and military interests. The offensive potential of ballistic missiles, especially when they may carry weapons of mass destruction, makes them a threat to deployed U.S. forces and regional stability. The relatively thin market for space launch and ballistic missile technologies is vulnerable to disruption by government interventions such as direct subsidies or inducements to sales (e.g., promising political favors for certain buyers). Economic considerations play a large role in national decisions to create a ballistic missile or space launch capability. Thus, stemming ballistic missile proliferation requires convincing countries that the economic as well as political and military benefits of having ballistic missile capabilities are not worth the costs. This is applicable to all three options.

Enhancing Global Partnerships with Other Spacefaring Nations Across the Spectrum of Economic, Political, and Security Issues. In this option, global partnerships to accomplish national security objectives would be commercial in nature rather than military but would be tempered by competitiveness issues. A common theme in all three options is the importance of economic factors in the effective exercise of military spacepower. Whether one calls it "raising the cost of entry" or "deterrence," a productive, competitive U.S. space industry can contribute to U.S. military objectives both by supporting U.S. forces and by shaping the global environment for space operations to the advantage of the United States. As with other commercial information industries, the key challenge for space-based information systems is to serve only authorized users. Both commercial and military space users are concerned with protecting the security of their investments, deciding how tasking requests are to be processed, and attracting the support of new users—whether they are customers or commanders. The United States can increase its national power if it can convince others that the U.S. government or U.S.-based firms are the most effective and reliable sources of meeting their needs for access to space and space-related information services. To do this, the United States must be seen as a reliable commercial partner whose national military and economic policies are balanced and mutually supportive to the greatest extent possible.

Accomplishing National Security Objectives: The Enhanced Strategy Option

Preserving Freedom of, Access to, and Use of Space. In this strategy, ensuring access to space would be the responsibility of the Air Force for the DoD, albeit supplemented primarily by commercial space launch vehicles. Achieving this objective also depends on maintaining an effective capability for space-based surveillance of the earth and of objects in space. Since this option envisions greater integration between the military and intelligence communities, conceivably there would be a greater emphasis on exploitation of available means from both communities to deal with rogue or Third World missile threats. Preservation of independent access to space by each space sector may place a high premium on the deployment of space control capabilities[11] by the military, as well as consideration of alternative survivability measures on commercial and civil as well as military space assets.

Maintaining the U.S. Position as the Major Economic, Political, Military, and Technological Power in Space. Since this option encompasses multiple players, it places a premium on multisector coordination and cooperation, guided by a coordinated interagency approach to policy development and implementation. As was shown in Table 3.1, the United States currently leads the world in capabilities and technologies across the board, except for space launch, in which Europe leads. Developing a cost-effective way to place payloads in orbit is fundamental to maintaining the U.S. lead in all other areas—and a necessary prerequisite to all aspects of spacepower. But it is also dependent on budgetary resources to enable the exploration of new technologies, on market demand, and on economic strategies such as having the U.S. government be an "anchor

---

[11]This could, for example, entail the employment of a strategic bomber like the B-2 in targeting satellite ground stations in a hostile theater of operations.

tenant" to the commercial launch industry.[12] In this option, the absence of a strong commercial launch industry capable of meeting all of the government's needs means that the DoD must devote precious resources to this element of spacepower, with the possible result that it cannot fund other systems or technologies that might have operational utility to the warfighter (e.g., space-based weapons) and at the same time enhance the U.S. position globally.

Deterring Threats to U.S. Interests in Space and Defeating Aggression If Deterrence Fails. Again, given the number of players involved, an integrated approach to supporting this objective is necessary. DoD must balance its interests against those of the commercial and civil players, but could possibly frame any potential issues in terms of military support to economic interests. Deterrence of potential threats to U.S. interests could also entail the purchase of both commercial satcom and high-resolution remote-sensing data of areas of interest to U.S. military operations. That purchase in turn could become a signal of intent and have a deterrent effect on potential aggressive actions by an adversary. Defeating those actions, however, should deterrence fail, would likely depend on using terrestrial military forces to protect U.S. space assets. In this case, CINCSPACE could request support from another CINC to protect ground stations located in a theater of conflict, for example.

Preventing the Spread of Weapons of Mass Destruction to Space. As with the Minimalist strategy option, stemming ballistic missile proliferation requires convincing other nations that the benefits of such capabilities are not worth the costs. However, this option has more means of enforcing that argument, in the form of space control and limited BMD capabilities in place. Coordinating information from civil and commercial imagery, exploiting commercial communications, and backing them with enhanced reconnaissance, ITW&AA, space control, and terrestrial forces, all in coordination with economic and political strategies, should help to achieve this objective.

Enhancing Global Partnerships with Other Spacefaring Nations Across the Spectrum of Economic, Political, and Security Issues. As in the Minimalist strategy option, "shaping the future battlefield" necessitates a coordinated and proactive approach by the DoD to ensure that it has access to civil, commercial, and international space systems. This has the added benefit of enhancing existing global partnerships with U.S. allies and friendly spacefaring nations, and possibly developing new partnerships based on mutual military and economic concerns. Since the Air Force is the primary developer and operator

---

[12]There has been extensive debate on how anchor tenant agreements should be constructed. Some argue that the government should be the first dollar in, providing the initial funding for a project that is of value to the government. Others argue that it should only be the last dollar in, adding its business to projects that can stand on their own commercially.

of U.S. military space systems in this option, it would bear the responsibility for articulating how the services could satisfy CINCSPACE and joint needs for operational support from space systems.

Accomplishing National Security Objectives: The Aerospace Force Strategy Option

Preserving Freedom of, Access to, and Use of Space. As this option includes strong space control and force application capabilities, coupled with a combination of commercial space launch and DoD quick-reaction launch capabilities, protection of critical space assets should be effective and access to commercial space capabilities should be ensured. A mix of DoD and commercial navigation, remote-sensing capabilities, commercial satcom, and government-provided weather and warning capabilities implies that the DoD has convinced U.S. commercial providers that they share common interests in safeguarding American industry's interests and potential opportunities. Backing that understanding is the enforcement aspect of the option: space control, force application, and ballistic missile defense, which should deter potential adversaries from undertaking any action in space or against U.S. and allied interests.

Maintaining the U.S. Position As the Major Economic, Political, Military, and Technological Power in Space. Implied in this option is a coordinated military, civil, and commercial approach to accomplishing this objective, guided by a strong interagency-developed policy and strategy. Assuming that the Air Force has evolved into the Aerospace Force, a focus on aerospace power to accomplish this objective and others would likely entail a multifaceted approach consisting of coordination among all sectors in U.S. space activities. The option thus offers the United States a stronger foundation and basis from which to strengthen its competitive position in the international, political, and economic environment.

Deterring Threats to U.S. Interests in Space and Defeating Aggression If Deterrence Fails. Dominance of the aerospace medium through the use of air- and space-transiting vehicles and weapons should support deterrence and, if deterrence fails, provide the U.S. military with the capabilities to defeat threats to U.S. and allied interests. This option will enable the National Command Authority and theater CINCs to have a greater range of decision options in crises or conflict. Those decision options may emphasize the use of aerospace power rather than naval or land power, assuming that, for purposes of discussion here, force structure choices among land, sea, air, and space forces were made to establish a robust aerospace force.

Preventing the Spread of Weapons of Mass Destruction to Space. As in the other two options, this important objective is more likely to be realized in this option, given the presence of ballistic missile defense and robust surveillance and warning capabilities coupled with a range of economic and commercial strategies. Since the merging of intelligence and military space sectors is implicit in this option, there should be a seamless approach to determining intent, assessing capabilities, and enforcing policy measures to prevent the spread of WMD to space. This seamless approach is also necessary to "shape the battlefield" in case U.S. forces become involved in crisis or warfare somewhere on the globe.

Enhancing Global Partnerships with Other Spacefaring Nations Across the Spectrum of Economic, Political, and Security Issues. The capabilities that this option includes would make it likely that alliance and coalition relationships could be fulfilled, and that U.S. and allied space-based and terrestrial assets could be protected. Fulfilling alliance relationships could entail joint partnership in developing new capabilities, or exploiting allied space capabilities in instances where U.S. capabilities are unavailable. Determining what contributions spacefaring allies could make to augment U.S. forces when engaged in pursuing alliance interests is required for this option (and probably the other two as well). To be kept in mind is the fact that nations operate in pursuit of their own vital interests, and a situation may occur in which the United States must rely on its spacefaring allies to fulfill U.S. national interests because U.S. space forces have been constrained operationally or politically. Therefore, procedural and operational arrangements need to be worked out early on that satisfy alliance goals and objectives as well as deter conflict.[13]

How the Options Might Support Operations Across the Spectrum of Conflict

Assessing the strategies described above yields very different organizational answers to questions of fulfilling the tasks and functions discussed in Chapters Two and Three. If the DoD is very dependent on outside sources for critical information, then it either cannot accomplish national security objectives or must support them in distinctly different ways from current practices. The Minimalist approach to exploiting space-based systems—a notion that is entirely conceivable in today's budgetary environment—necessitates more attention to economic factors underlying military spacepower, potentially through expanded military involvement in governmental regulatory and licensing activities, through developing innovative approaches to incentivizing industry to invest in the DoD market, and through encouraging greater exploitation of allied or friendly space capabilities. All these activities require years of DoD involve-

---

[13]Johnson (1990), pp. 2–3.

ment in nontraditional fora. At the other end of the spectrum, the Aerospace Force option does not depend on commercial capabilities, but it may still require some of the same activities to shape the battlefield in advance of potential military operations. Furthermore, because of trends in the other space sectors, DoD's implementation of this option may not be possible unless it actively considers and incorporates these nontraditional activities outside its immediate purview.

Referring to Tables 2.1 to 2.3, where we address space support for a range of operations (for illustration purposes only), we will describe what it means to perform a particular mission having one of the three options in place. The examples we have selected are purely illustrative and not evaluative. Each strategy option is described as to how it would support first, the type of operation, and second, a specific operational objective.

Space Support for Peacekeeping/Humanitarian Operations: Establish and Defend Safe Areas. As described in Chapter Two, operational tasks to support this objective might include the monitoring of intermediate range ballistic missile (IRBM) launches in areas of crisis; ensuring secure communications between the U.S. and coalition forces; denying infiltration in regions of concern; establishing accurate boundaries of safe areas; ensuring freedom of movement to, from, and in space; and monitoring potential threats to safe areas. The Minimalist strategy option's strength and effectiveness here lie in providing ITW&AA and a measure of space control, supported by reconnaissance capabilities and good relationships with civil relief and commercial space agencies and industry to provide communications, navigation, and weather monitoring over the disaster area. However, it would likely be dependent on terrestrial military forces, whether they are U.S. or local government forces, to defend those safe areas and ensure safe passage of people, food, and medicines. The Enhanced strategy option would rely on existing space forces, supplemented by civil and commercial capabilities and by terrestrial forces, largely for site defense. Its limited ballistic missile defense capability would be used to deal with potential missile threats to the safe areas. Finally, the Aerospace Force strategy option might conceivably employ space-to-ground weapons to ensure defense of the safe areas from a range of threats. It would still be dependent on commercial satcom and commercial space launch, as well as a mix of DoD and commercial navigation systems and imagery of the disaster area.

Space Support for Peacekeeping/Humanitarian Operations: Conduct Disaster Relief. The functions contributing to this operational objective include monitoring weather, establishing communication links in disaster areas, and assessing the local terrain for rescue operations that might be conducted. The Minimalist strategy option would be heavily dependent on commercially available imagery and communications as well as environmental monitoring from

NOAA. Military space forces would likely not contribute much to this situation, in contrast to the Enhanced and Aerospace Force strategy options, which would have more robust capabilities and be more likely to have an interagency process in place to respond to requests for support from federal, state, and local authorities (for domestic disasters) and foreign governments (international disasters).

Space Support for Crisis Operations: Deter Aggressive Actions by Belligerents. Operational tasks using space-based systems to help deter belligerent actions might include assessing sea state effects on enemy naval activities; conducting surveillance of troop movements in theater; neutralizing hostile artillery; and initiating preparatory BMD actions as both a signal of intent and as necessary to ensure adequate defense if deterrence fails. Again, the Minimalist option relies heavily on those national-security-related functions for which there are military forces—ITW&AA, reconnaissance, ballistic missile defense, and space control—and on commercially available imagery, navigation/positioning information, and satcom. Deterrence might also include economic actions far in advance of a potential crisis, such as attempting to restrict access by a belligerent to commercially available remote-sensing data; however, this action carries other political and foreign policy risks, which may be unacceptable. Because more agencies are involved in performing similar functions, the Enhanced strategy option may provide redundancy of information from space-based sensors to contribute to deterring adversaries. Finally, the Aerospace Force strategy option can contribute to deterrence through a coordinated approach (with commercial providers) of imagery and satcom, and the enforcement measures of ballistic missile defense, force application capabilities, and space control. All three options would also include close coordination with allies in the region where the United States is attempting to deter. Protection of ground-based commercial space interests is also necessary, placing a heavy premium—particularly on the Minimalist strategy option—of working with alliance governments, which would have to see it in their best interests to protect U.S. systems and ground sites.

Space Support for Regional Conflict (MRC Level): Halt/Evict Invading Armies. As discussed above, the operational objective of halting or evicting invading armies in support of theater or regional conflict could perhaps be met by the following operational tasks: ensuring adequate and secure communications for response; determining routes of attack; and maintaining the position location of U.S. and allied forces. These tasks can be accomplished by all three options, although obviously in different ways, whether by relying on nonmilitary sources or on military-owned systems for the information. They also involve coordination with other military forces, both U.S. and allied. In the Minimalist option, the United States may have to depend on allied surveillance capabilities to monitor the theater in question; as in other instances, this requires extensive forethought and preparation ahead through policy and regulatory initiatives. Although the other two strategy options are less dependent on those considerations, they are still important factors to be considered in developing military response options.

THE GROWING IMPORTANCE OF TRUST AND SPACE LITERACY

Implicit in our discussion throughout this study is the notion of cross-cultural cooperation among varying space sectors, each with different goals, objectives, and interests. As discussed in other works, the key to accomplishing cross-cultural cooperation lies in trust and space literacy. The importance of both is described below:

> While trust among the Services is a key factor underlying the difficulty in assigning functional and organizational responsibilities for space, trust at other levels is also important. The extent of space literacy among all players in space is linked to trust, since having little or no trust in another organization's ability or willingness to address one's needs for space-derived information somewhat depends on one's own understanding of what is involved in meeting that need. Only by gaining a certain level of self-confidence, familiarity, and competency in understanding space operations, programs, and systems can the warfighter feel comfortable with allowing a single organization to take responsibility for meeting his needs....[14]

This observation is applicable not only to the services, within individual services, and within the DoD as a whole, but also to the other sectors of U.S. space activities and at the international level. Space literacy is intrinsically linked with trust and entails a certain level of understanding about the advantages, disadvantages, and kinds of capabilities that space systems offer. To accomplish national security objectives and conduct military operations that use space-derived information, warfighters need to be sufficiently knowledgeable about space operations to articulate requirements and needs to the service that provides them. Concomitantly, they need to be able to explain those needs to the commercial space world that will likely develop the technologies and systems to satisfy those needs.

Applied to the strategy options discussed in this chapter, the Minimalist option is clearly very dependent on having a trusting and cooperative relationship with the various sectors with which it interacts; the Aerospace Force option is less dependent on but should be no less interested in similar relationships. In fact, developing and implementing the kinds of economic and military strategies we

---

[14]See Johnson et al. (1995), p. 79.

have described throughout this document cannot be accomplished without a firm foundation of trust among all the players, and an understanding about accountability and responsibility. Specific objectives and criteria to measure progress toward achieving greater trust and space literacy should be made part of the implementation portion of these strategies, or at the very least be an integral part of the interagency process by which the strategies are developed.[15]

---

[15] See Johnson et al. (1995), especially Section 4.

Chapter Five

UPDATES SINCE 1994

The purpose of this chapter is to provide a brief update on important events in the evolution of thinking about spacepower since the bulk of this study was written in 1994. A new National Space Policy has been released, and thinking about military space operations has grown with the release of military vision statements at various levels. Perhaps most important, commercial space activities have continued to grow dramatically, providing both new opportunities and challenges to the U.S. military.

GROWTH OF SPACE COMMERCE

A report on "The State of the Space Industry" estimated that 1996 revenues for the global space industry exceeded $76 billion.[1] The study divided the space industry into four sectors:

- Infrastructure—e.g., ground systems, satellites, and launch vehicles. 1996 revenues of $47 billion;

- Telecommunications—fixed and mobile satellite services, direct-to-home television. 1996 revenues of $9 billion. Indirect revenues from satellite cable distribution and telephony were about $13 billion;

- Emerging applications—e.g., remote sensing, GPS applications, and geographical information systems accounted for about $4 billion in 1996 revenues;

- Support services—e.g., financial services, insurance, consulting, and publishing directly related to the space industry were estimated to generate $3 billion in revenues.

---

[1] Space Vest (1997).

Worldwide, space industries were estimated to employ over 800,000 people, expanding at the rate of 40,000 jobs per year. Although governments were the major source of space revenue, high industry growth rates were being driven by commercial activity and changing the relative mix of public and private revenue sources for virtually all space firms. As discussed above, the implications of this growth affect all areas of space operations. For example, new mobile satellite communication services create demand for more launch services, which increases the ability to raise private capital for improving the performance and capacity of space launchers. Increasing commercial use of GPS and remote-sensing data helps drive down the price of related equipment and services, which can then be exploited for military purposes. On the other hand, these same space-based information systems can also be used by adversaries, and thus military and diplomatic countermeasures must be developed (and are currently in work).

U.S. POLICY AND STRATEGY

At the same time that commercial space revenues have been increasing, DoD budgets have been declining. According to Secretary of Defense William Cohen:

> Since 1985, America has responded to vast global changes by reducing its defense budget by some 38 percent, its force structure by 33 percent, and its procurement programs by 63 percent. Today, the DoD budget is $250 billion, 15 percent of the national budget, and an estimated 3.2 percent of our gross national product.[2]

The defense budget is expected to be fairly stable for several years, barring a major crisis. How the defense budget will be allocated is a topic of continuing debate. The National Military Strategy states that U.S. military objectives are to:

> ... promote Peace and Stability and when necessary, to defeat Adversaries that threaten the United States, our interests, or our allies. U.S. Armed Forces advance national security by applying military power to Shape the international environment and Respond to the full spectrum of crises, while we Prepare Now for an uncertain future.[3]

The strategy does not mention space per se, but it is consistent with our view that the international environment for spacepower *can* be shaped and that preparation now can ensure future military options even in a constrained bud-

---

[2]Cohen (1997). In 1985, the DoD budget was about $400 billion in 1997 dollars and represented 7 percent of the U.S. gross national product.

[3]Office of the Joint Chiefs of Staff (1997). Capitalization in original.

get environment. More specifically, the strategy cites *Joint Vision 2010* as the conceptual template for joint operations and future warfighting. *Joint Vision 2010* (JV 2010) was released by the Chairman of the Joint Chiefs of Staff in April 1996 to provide guidance on how the U.S. military would "achieve dominance across the range of military operations through the application of new operational concepts . . . ."[4] Four operational concepts were defined, all based on information superiority and technological innovations:

- Dominant maneuver,
- Precision engagement,
- Full-dimensional protection, and
- Focused logistics.

Again, space is not specifically mentioned, but the ability to implement each concept depends on the effectiveness of space systems. Space systems enable necessary characteristics such as situational awareness, responsive targeting, and combat identification. The emphasis of current space systems is to provide information from and through space, but there is increasing interest in space control and force application mission areas. Even with the end of the Soviet Union, the interest in space control is understandable given the increasing importance of both civilian and military space systems to the United States. Full-dimensional protection will require some form of ballistic missile and cruise missile defense, and again, space systems will be needed. Applying force from space, however, continues to be controversial from many technical, economic, and political perspectives. Nonetheless, the capabilities that space provides today are increasingly recognized as vital for the future effectiveness of U.S. forces.

A new National Space Policy was released on September 19, 1996.[5] It provides top-level policy goals and guidelines for civil, commercial, and national security (military and intelligence) space activities. Much of the guidance on military space matters is a continuation of past policies with updates on the pursuit of ballistic missile defenses and relations between military and intelligence space activities. Most notably, the DoD is directed to maintain the capability to execute the mission areas of space support, force enhancement, space control, and force application. No detailed guidance is provided for force application, but with respect to space control, the policy specifies that:

---

[4]*Joint Vision 2010* (1996).
[5]The White House (1996).

> Consistent with treaty obligations, the United States will develop, operate, and maintain space control capabilities to ensure freedom of action in space and, if directed, deny such freedom of action to adversaries. These capabilities may also be enhanced by diplomatic, legal, or military measures to preclude an adversary's hostile use of space systems and services.

Exactly how these capabilities are to be made operational is still a subject of intense debate, however, as are various proposals for antisatellite weapons and arms control agreements.

The next section will briefly discuss the evolution of the U.S. Space Command and its vision of the future as a primary example of current U.S. thinking about the role of spacepower in warfare.

U.S. SPACE COMMAND

In light of the many changes that had occurred since the creation of U.S. Space Command (USSPACECOM) in 1985, the command conducted a "revalidation" study of its mission and organization in 1996. The results of the study were briefed to the Secretary of Defense in November 1996.[6] The proposal to create a unified Space Command was initially made in 1983 in response to growing strategic and space threat capabilities (primarily Soviet), and the need for an operational focal point for the many fragmented space activities of the services and the Strategic Defense Initiative (SDI) to create ballistic missile defenses.[7] SDI in particular provided an image of defensive weapons in space and the need for these weapons to be under military command. It was also thought that the existing fragmented arrangements for space systems would not function effectively in wartime and new space warfighting concepts were needed.

In 1986, the Goldwater-Nichols DoD Reorganization Act (also known as Title 10 from its location in the U.S. Code) was passed. Title 10 dictates a separation between CINC responsibilities for warfighting and service functions to provide forces to the CINCs (e.g., organize, train, and equip). Responsibilities and authorities given to CINCs were increased: They were now to "direct, organize, and employ" forces for all aspects of military operations, joint training, and logistics. This reinforced the decision to create a separate command for military space operations.

The Persian Gulf War during 1990–1991 raised both public and military appreciation of the combat support provided by space systems. Attention to naviga-

---

[6] *Mission Revalidation Brief* (1996).

[7] The Air Force Space Command was established in 1982. The Army and Navy Space Commands were established in 1988 and 1983, respectively.

tion, communications, weather, and missile warning functions during the war demonstrated that space systems were useful at the tactical level and not confined to strategic roles (e.g., nuclear operations and intelligence). In 1992, these "force enhancement" functions were assigned to the U.S. Space Command as a mission area under the Unified Command Plan (UCP). This was in addition to earlier-assigned functions such as space control and space support (e.g., launch). In 1993, "force application" was added as well, to allow for employing weapons from, in, and to space.

The revalidation study concluded that the rationale for having a Space Command had endured and grown, even into the Ballistic Missile Defense Organization (BMDO), the decline of the Soviet threat, and the redirection of the Strategic Defense Initiative. The focus of space warfighting had shifted to include operational and tactical, as well as strategic, applications. A renewed emphasis on theater missile defense was given in the latest national space policy. Technological changes since 1985, most notably in information technology, had created a wide range of new tactical space capabilities. Finally, the growth of commercial space activities meant that nonmilitary space systems were even more important to the U.S. economy and global infrastructure. This meant an investment that might need protection in a future conflict.

In 1996, the U.S. Space Command published a vision statement that saw itself as: "Dominating the space dimension of military operations to protect U.S. interests and investment. Integrating Space Forces into warfighting capabilities across the spectrum of conflict."[8] Like JV 2010, the *USSPACECOM Vision* described four operational concepts. These were:

- Control of space (e.g., assured access, protection, negation),
- Global engagement (e.g., global surveillance, precision strike),
- Full force integration (e.g., coalition interoperability, training), and
- Global partnerships (with civil, commercial, international space capabilities).

These operational concepts are seen as supportive of, and consistent with, the JV 2010 operational concepts. For example, control of space supports concepts such as dominant maneuver, precision engagement, and full-dimensional protection. Global engagement tracks with the Joint Vision concept of precision engagement. The concept of creating global partnerships is especially appropriate for space, because the DoD can no longer afford to rely solely on unique military capabilities. Partnerships with civil agencies, companies, and

---

[8] *USSPACECOM Vision* (1996).

international organizations may allow for sharing costs and risks for military advantage. On the other hand, certain functions such as missile warning and the use of weapons can and should be retained as unique "core capabilities" because of the importance they have in supporting critical military operations and warfighting.

The U.S. Space Command produced a "vision implementation plan" based on the four operational concepts in its vision statement. The commander's intent for the plan states that:

> This Command will defend national interests as a supporting or supported CINC with space forces that are fully integrated with land, sea, and air forces . . . USSPACECOM capabilities will be augmented with commercial, civil, and international space systems. The Vision Implementation Plan will provide USSPACECOM and its components a roadmap and measuring tool to attain the required capabilities for 2020.[9]

The plan seeks to create a "roadmap" for each of the four operational concepts, resulting in changes to existing doctrine, organization, and training as well as new capabilities and technologies.

Of the four operational concepts, global partnerships is likely to be the most difficult and important one to implement. It requires that the military find common ground with very different institutions and cultures that may have little to no interest in national security concerns. The role of military spacepower has been in transition for many years, and its role in tactical operations across the spectrum of conflict has become more routine and accepted. The idea of spacepower as inclusive of nonmilitary space capabilities is less well known. The DoD is increasingly in a minority role with respect to the technical and economic development of new space capabilities. However, the global environment (or future battlespace) for space systems can be shaped by DoD actions today. Examples of proactive shaping opportunities include promoting GPS as a global standard, ensuring that U.S. firms dominate commercial remote-sensing and mobile communications markets, and regaining global market share for U.S. launch vehicles while undermining economic incentives for new entrants (and the risk of missile proliferation). Each of these opportunities can be influenced by how military space requirements are defined and implemented, e.g., by whether the requirements seek to leverage commercial forces or ignore them.

Under Title 10, USSPACECOM does not have the "organize, train, and equip" charter and has limited resources for implementing broader national security strategies. However, it does have a responsibility to communicate the warfight-

---

[9] *USSPACE Vision Implementation Plan* (1997).

er's needs for space-based information through its components to the service departments that develop the capabilities to fulfill the CINC's needs. USSPACECOM should be able to pursue partnership opportunities that improve operational readiness through information exchange, build relations that may be beneficial in wartime, and improve its ability to advise others on how to proactively shape (if not direct) the coming generation of civil and commercial space systems. But it should also encourage its components (Air Force Space Command, Army Space Command, and Navy Space Command) to "build bridges" to their respective services, primarily through the requirements process.

WORKS IN PROGRESS

The 1994–1997 period saw a considerable level of effort directed at new thinking about the military uses of space. The most significant intellectual developments have been in defining a National Space Policy, a National Military Strategy, *Joint Vision 2010*, and a *USSPACECOM Vision*. Unfortunately, two important documents, in work in 1994, have yet to be completed and released. The first is a DoD Space Policy that would provide more detailed guidance for the implementation of the military and intelligence aspects of National Space Policy. In particular, policy guidance is needed on space control and force application missions. The second is Joint Pub 3-14, "Joint Doctrine, Tactics, Techniques, and Procedures for Space Operations." Joint doctrine is intended to serve as authoritative guidance for all U.S. forces and provide a focus for system applications and technology. JV 2010 describes joint doctrine as "the foundation that fundamentally shapes the way we think about and train for joint military operations." Drafts of Joint Pub 3-14 have addressed topics such as principles of joint space operations, the command and control of joint space operations, and planning and support procedures (e.g., processes for requesting space combat support). The military space community has not yet come to grips with how to apply existing policy and doctrine to the specific challenges of using spacepower.[10] In an effort to specifically address the theoretical basis for military spacepower, CINCSPACE commissioned a study in March 1997 to create a theory of spacepower. Expected to be completed in late 1998, the study is intended to be a theoretical treatment of spacepower analogous to the classic military writings of Clausewitz, Mahan, and Douhet.

The next chapter will discuss some of the most significant, and difficult, policy and doctrinal challenges in exploiting the potential of spacepower.

---

[10]Air Force space doctrine has also languished, primarily because of the lengthy approval process and attention to the development of a corporate vision ("global engagement") and doctrine (AFDD-1).

Chapter Six

CHALLENGES FOR THE FUTURE

There are many difficulties in effectively exploiting spacepower for national security objectives. The most serious of these difficulties are not technological or even fiscal, but operational, doctrinal, and organizational. For example:

- How should space capabilities be integrated into other military operations?
- What should be the core competencies of military space forces?
- What priority should information warfare and space-based weapons have in developing military space capabilities?

Air and space integration raises a number of significant organizational issues that reflect important choices about how spacepower is to be employed. In part, these choices are similar to earlier debates about the role of air forces, such as whether and when they should be under unified or independent commands, what priority should be accorded to air operations, and what relation they should have to land- and sea-based commands. This is part of the continuing debate over centralization and decentralization, e.g., the distinction between having "unity of command" and achieving "unity of effort."[1] In the case of space operations, joint doctrine debates continue over questions such as:

- Should there be a "space" JFACC, or even a Joint Force *Aerospace* Component Commander? What is the scope of his responsibility and authority?
- How should an air-space tasking order be constructed?
- Should space be declared a regional area of responsibility (AOR)?

Currently these questions are being addressed not only at USSPACECOM but also at the component level. The commander of the Air Force component to

---

[1] Winnefeld and Johnson (1993).

USSPACECOM, the Fourteenth Air Force, dual-hatted as the commander of AFSPACE forces (COMAFSPACE), at Vandenberg AFB in California, has articulated an approach to the command and control of AFSPACE forces and has established a Space Operations Center (SOC) at Vandenberg. The intent is to ensure the most effective use of Air Force space forces in support of CINCSPACE, other unified commanders, and their JFACCs.[2] This and other efforts represent the beginnings of truly operationalizing and normalizing space operations consistent with mainstream military operations.

Documents such as *Joint Vision 2010* and the *USSPACECOM Vision* recognize that modern warfare requires fighting as a joint, integrated team. Because space forces are the newest component of national military power, the Air Force has less experience with their integration than land, sea, and air forces. As a result of decisions made at the CORONA (October 1996) meeting of senior Air Force leadership, the Air Force stated its intention to transition from an air force to an "air and space force" to an eventual "space and air force." Accomplishing this will require major institutional changes in Air Force operations, organizations, education and training, career assignments, and budgets. RAND is currently completing a separate study on the integration of space into Air Force operations.[3] Further work will likely be needed on space integration in joint operations with the other services and combined arms operations with other countries.

The integration of space forces with other military forces raises more basic questions: first, whether space is a region of vital interest to the United States, and second, what should be the "core competencies" of a military space force that would operate in that region. Establishing "space" as an AOR of CINCSPACE is related but not the entire issue. As we have discussed throughout this document, a wide variety of both military and nonmilitary activities are conducted by a large number of different countries and organizations in the environment of space. No hostile power dominates the environment of space or the activities conducted in space, and the unique characteristics of spaceflight mean that objects are continually in transit over most, if not all, regions of the earth. Operations conducted in space provide the advantages of both a global perspective and a global presence for the nation (or company, for that matter) that puts a spacecraft or a manned vehicle in orbit. Under certain circumstances, however, that activity could be perceived as either stabilizing or destabilizing by the country over which that vehicle is transiting, depending on that country's own investment in space operations.

---

[2]*Command and Control of AFSPACE Forces: A White Paper Prepared to Articulate the Vision of COMAFSPACE* (1997).

[3]Johnson et al. (1997).

Given the expansion of U.S. and international commercial investment in both communication systems and the information from space-based sensors, great incentives exist for preserving freedom of access to space and to the data transmitted through space for U.S. national and economic security interests. Declaring the environment of space to be a region of vital national interest could be justified on the basis of the criteria defined in *A National Security Strategy for a New Century* (1997). The National Security Strategy defined three categories of national interests: vital, important, and humanitarian. "Vital interests" are "those of broad, overriding importance to the survival, safety and vitality of our nation. Among these are the physical security of our territory and that of our allies, the safety of our citizens, and our economic well-being. We will do whatever it takes to defend these interests, including—when necessary—using our military might unilaterally and decisively."[4]

The above criteria (physical security of the nation and our allies, public safety, and national economic well-being), traditionally applied to territorial regions, also directly apply to the region of space, since many capabilities upon which we have come to depend can be provided effectively only from that region. For example, space-based early warning of ballistic missile attacks is absolutely essential to the civil defense of our population centers and, as we saw in Desert Storm, those of our allies. Space-based communications, navigation, intelligence, precision targeting, and many other military capabilities are essential to the military operations required to defend our other national interests. Satellite-derived precision navigation for commercial airlines and warning of severe weather, such as threatening hurricanes, help ensure the safety of our population. The $1.5 trillion (and growing) spinoff benefits to our economy from commercial space activities enhance the economic well-being of our nation, not to mention the less quantifiable advantages afforded commercial enterprises through space products and services such as communications, navigation, weather, and remote sensing. One must conclude that the region of space, which enables all of these critically important capabilities, is a region of vital national interest, and the loss of access to this region could be devastating to the nation as a whole.

Deciding how to defend that vital national interest, through a combination of tools (diplomatic, economic, military, etc.), then leads to consideration of the kinds of military forces and capabilities required. Related to this discussion is the question of just what the "core competencies" of a military space force should be. This question is very challenging and is still open for debate. Unlike

---

[4]The White House (1997), p. 9. "Important national interests" are defined as interests that "do not affect our national survival, but they do affect our national well-being and the character of the world in which we live." "Humanitarian interests" applies to actions undertaken to respond to natural or manmade disasters or gross violations of human rights.

other types of military systems (e.g., tanks, bombers, and aircraft carriers), today's military space systems are not generally thought of as weapons per se, but are considered as force enhancement capabilities. That is, they do not destroy targets directly but enable destruction through the information services they provide. Furthermore, many military and commercial space activities (e.g., navigation, remote sensing, weather, communications, and launch) are dual-use and not obviously military functions. Consequently, the potential use of these dual-use systems for enforcing a declaration that the region of space is a vital national interest raises a number of other questions that warrant careful consideration.

The diverse range of existing and projected commercial space capabilities raises the question of what the military *should* do in space for itself, not what *can* it do. As discussed in Chapter Three (see Figure 3.1), the spectrum of space-based activities ranges from those that are purely military to those that are almost entirely commercial. In deciding which activities should be retained under military control rather than being "outsourced," the following five criteria may be helpful:

- Activities where military persons (i.e., subject to the Uniform Code of Military Justice) are required to be in control, as in the employment of weapons and potential exposure to armed conflict;

- Activities that are so critical to national security that they must remain responsive to the National Command Authority;

- Activities where military control is cost-effective or the only practical option compared to commercial or civilian control;

- Activities where military control, on balance, helps shape the battlespace in a manner favorable to the national interest of the United States; and

- Activities where military control is necessary for the preservation of skills to meet unique requirements.

The convergence of the DMSP and NOAA-POES weather satellite programs is an example of where exclusive military control was not cost-effective when civil and military requirements could be met by a single system. GPS is another example of a space system that meets both military and nonmilitary requirements but is under military control. The DoD role in GPS is made necessary by the importance of GPS to critical national infrastructures as well as military operations. Although some foreign governments are uncomfortable with DoD stewardship of GPS, this role is seen as a reassuring sign of stability and quality by the vast majority of international public and private users of GPS. The DoD role helps deter the emergence of competing systems and helps promote GPS as a global standard that is beneficial to the United States and its allies.

Space launch is an area where DoD already depends on others for its capability. In the future, DoD may have its own dedicated fleet of reusable military space planes, but such autonomous access to space does not exist today. This has raised several concerns, such as how the DoD would be able to ensure an ability to "surge" and "reconstitute" space systems during a conflict. Today, the United States has few capabilities in these areas, particularly for large payloads. Another concern is whether some level of space launch skills should be retained in the uniformed services. On one hand, it seems ironic that access to space would not be considered a "core competency" of a military space force. On the other hand, the lack of a truly warfighting launch capability seems to have made the retention of skills for current launch vehicles increasingly difficult to justify. Peacetime launch skills are resident in industry, so there is little motivation to duplicate such skills in a military force.

With the exception of weapons targeting, control, and release, it seems that all space activities would be potential candidates for civil/military partnerships and even outsourcing. As discussed above, USSPACECOM has made the creation of global partnerships one of its four operational concepts in support of its vision statement. The National Space Policy also talks about the importance of international cooperation and even cooperation with state and local governments in achieving national objectives. What criteria, then, should guide the definition of global partnerships and associated organizational responsibilities? CINCSPACE would focus on long-term planning issues, whereas its components would deal with operational issues, and the military services would be developing and procuring capabilities to meet CINC and operational requirements. As noted above, the division of responsibilities under Title 10 places budget and responsibility for most potential partnership opportunities with the services (i.e., who provide forces to the CINCs), not the CINC himself. However, the CINC plays a very important role in helping to shape regional issues through interactions with other nations' political and military leaders, and by providing space expertise to other U.S. CINCs who may not have that capability resident in their theaters. Thus, he can, and should, explore bilateral and coalition partnerships that enhance warfighting capabilities. But any space-related partnering relationships he undertakes need to be coordinated with the geographical CINCs and their own efforts at regional partnerships for other military reasons.[5] Clarification of organizational responsibilities and authorities on

---

[5]On January 29, 1998, the President signed a new UCP outlining CINC responsibilities. CINCSPACE was given new authorities and responsibilities to include:

1. Serving as the single point of contact for military space operational matters,

2. In coordination with the Joint Staff and other CINCs, providing military representation to U.S. national, commercial, and international agencies for matters related to military space operations unless otherwise directed by the Secretary of Defense,

global partnership efforts and other matters could occur through the formulation and implementation of a national strategy for space and the development of a clearly defined interagency process.

Decisions on how space capabilities should be integrated with more traditional military capabilities, and which functions should be performed by the services themselves, are continually influenced by competing visions of what is meant by spacepower. We have used the term in a very broad and general way in this report. Nonetheless, others see spacepower as requiring the existence of space-based weapons capable of space-to-ground or space-to-space attack. Others, who may be opposed to such weapons on practical or ideological grounds, see spacepower as synonymous with using space systems to acquire or transmit information in support of more traditional forces. Although there is general agreement on the need to better integrate space operations into mainstream military operations, there is less agreement on what should be the top priority for acquiring new military space capabilities. There are various debates over proposals such as space-based lasers, reusable military space planes, new intelligence systems, and theater missile defenses.

The current high cost of access to space has meant that only the most valuable activities have justified the cost of a launch. In the commercial world, this has meant that the only profitable product from space has no mass, i.e., it is information, not manufactured products or tourists. Similarly, the most valuable military space activities today are those that handle information, not the movement of troops or weapons into space. If a low-cost military space plane were developed, it could have a major effect on the conduct of space operations in all missions areas (i.e., space support, space control, force enhancement, and force application). A responsive space plane could enable capabilities that do not exist today, such as launch surge and reconstitution and space fire support.

The most important policy and doctrine development for future U.S. military space capabilities is not likely to be from space-based lasers or even a reusable launch vehicle, however, but from space-based information technologies. This is especially true with regard to the development of operational concepts and

---

3. In coordination with geographic CINCs' security assistance activities, planning and implementing security assistance relating to military space activities,

4. Coordinating and conducting space campaign planning through the joint planning process in support of the National Military Strategy, and

5. Providing the military point of contact for countering the proliferation of weapons of mass destruction in space.

Planning for and implementation of these new responsibilities are under way as this document goes to print.

organizational structures that ensure effective employment of those information technologies. These technologies are being driven by private investments, spectrum allocations, commercial standards, changing market shares of competing firms, and international trade rules. DoD is not currently a major player in these issues and thus misses opportunities to shape the form of overall U.S. spacepower.

Developing space-based weapons may be easier and more attractive for the DoD than exploiting space-based information systems or even air-space integration. The reason is that, despite political and technical controversies, weapons and weapons-carrying platforms present few doctrinal and organizational issues. A space fighter is easier to explain conceptually than the process for making international spectrum allocations. Yet the latter may be more militarily significant to the implementation of *Joint Vision 2010* and the National Military Strategy. The military space community that emerged from the Cold War and the Persian Gulf War faces a difficult choice ahead. One path deals with difficult, often intangible issues in exploiting the information superiority that space systems can confer. The other seeks to develop exotic-sounding but culturally familiar space weapons and platforms but at the risk of diverting budget and institutional attention from more immediate problems.

It is certainly possible to imagine a future force that provides both information dominance and combat power from space. It is difficult to imagine being able to develop both aspects of such a force simultaneously. The limitations are not just budgetary and technological but involve organizational inertia and the intense competition for the time and focus of senior leaders and policymakers. In our view, priority should go to developing enduring space doctrine and exploiting information technologies. An aggressive program to develop new space launch capabilities would be helpful to the nation as a whole, with military funding of any unique requirements (e.g., survivability and rapid reconstitution). Space control is important for the protection of increasing private sector space capabilities (which may also be needed by the DoD). Aside from ballistic missile defenses, it is more difficult to justify the development of space-based weapons at this time but the option should be maintained through ongoing R&D activities.

CHOOSING A STRATEGIC DIRECTION

Each of the conceptual options discussed above—Minimalist, Enhanced, and Aerospace Force—represents a different approach to the conduct of space operations. Each can, and should, be compliant with JV 2010, National Space Policy, National Military Strategy, and other top-level documents, but their approaches to space operations represent different strategic choices in the exploitation of spacepower.

Minimalist: The military use of spacepower is highly dependent on external relationships and partnerships. Integration with other military operations depends on organizations outside the military chain of command.

Enhanced: The military use of spacepower is highly integrated with other forms of military power. External relationships and partnerships are important but not critical to core military capabilities.

Aerospace Force: Military spacepower is exercised through an independent service and fully capable of being exercised separately from other military forces. Actual military operations are most likely joint and combined and may employ external relationships, but this is not required.

Each option has different key challenges for DoD. In the Minimalist case, it is the establishment of dependable contractual and institutional relationships that enable the accomplishment of military missions with few indigenous capabilities. In the Enhanced case, it is the integration of space capabilities within USSPACECOM and in joint operations with all the services. Lastly, the challenge for the Aerospace Force is to demonstrate the existence of appropriate threats and a suitable doctrine to justify the necessary budget increases and organizational changes.

Choosing a strategic direction for the development of spacepower is hampered by uncertainty over the broader national security environment in which DoD will have to operate in the years ahead. In particular, there is a limited understanding of the effect of commercial activities and foreign industrial policies on the shape of future space and information battlespace. The ability of the U.S. government generally (not just DoD) to proactively shape the battlespace environment is limited by multiple, fragmented organizations. There tend to be many people working on system architectures and various technical challenges, but few people who concern themselves with competing commercial standards, spectrum licenses, and the international competitiveness of firms in the space industrial base. If the United States is to fully benefit from emerging opportunities in spacepower, it will have to:

- Expand its definition of spacepower to include nonmilitary space capabilities;

- Aggressively pursue the integration of space with other forms of military power;

- Identify and protect space-based functions that are critical to the nation as a whole (not being limited to military missions); and

- Work with nonmilitary organizations to shape the future battlespace for space operations, beginning with space-based information technologies.

Each of the three strategic options discussed in this report represents different outcomes for the core competencies of military space forces and different assumptions about reliance on external relations and partners. As stated above, partnering opportunities should be evaluated not only on how they benefit U.S. military space capabilities but on how they shape the international security environment to the advantage of the United States. This requires a broader outlook than traditional CINC or even SecDef responsibilities and may conflict with other DoD objectives unrelated to space operations, as well as with other federal government agencies' roles and responsibilities. Resolving these conflicts will likely require extensive interagency work and would benefit from the re-creation of a National Space Council or the strengthening of current national space policy mechanisms in the National Science and Technology Council.[6]

BOTTOM LINE

U.S. spacepower comprises national capabilities, not just the military space capabilities of the Department of Defense. The continued growth of the commercial space sector, particularly in information technologies, is therefore creating new national options for the exercise of spacepower for military, economic, and political objectives. At a time when the DoD budget is under severe downward pressure and the military space community is seeking to meet the needs of conventional warfighters, the use of commercial space systems holds the potential for enhancing U.S. military capabilities.

Commercial space systems are here to stay, both in the United States and elsewhere in the world. Regardless of the outcome of roles and missions debates, reorganization and management initiatives, or defense force structure reviews, the warfighter will need to know more about exploiting, partnering with, and countering space capabilities outside the military's more traditional focus, i.e., the commercial, civil, and international space sectors. This requires a proactive approach by DoD and the services to develop relationships with commercial space firms and create mechanisms to ensure that commercial space systems are not used in a manner hostile to U.S. forces. If U.S. industry is predominant in the commercial space sector, such relationships will be easier to establish and verify, enabling DoD to shape the future battlespace environment, gain maximum leverage from constrained resources, and deter the development of potentially hostile space forces.

---

[6]Strengthening the National Security Council's space functions is another option, but key agencies, such as the Department of Commerce, are not formal members of the NSC.

Finally, enhancing U.S. spacepower and providing new options for accomplishing national security objectives do not require a space "czar" or massive new expenditures. However, they do require a deliberate effort at coordination and communication between representatives of the nation's military, economic, and political interests in space, built on a strong foundation of trust, literacy, and cooperation. Only then can we understand the extent to which spacepower will influence the implementation of national security strategy and the conduct of future military operations in the context of exercising national power in a dynamic strategic environment.

BIBLIOGRAPHY

Aerospace Corporation, *Future Spacelift Requirements Study: Summary*, El Segundo, California, September 11, 1997.

Air Force Basic Doctrine, Air Force Doctrine Document 1, U.S. Air Force Headquarters, Washington, D.C., September 1997.

"Acquisition Reform Success Stories," http://www.safaq.hq.af.mil

Bedrosian, Edward, and Gaylord Huth, *Commercial Mobile Satellite Systems: The Potential for Tactical Military Applications*, RAND briefing, November 1994.

Berner, Lanphier & Associates, *Final Report to the Integrated Technology Applications Group on Communications Satellite and Space Launch Vehicle Technologies*, U.S. Department of Commerce, Office of Space Commerce, Washington, D.C., 1992.

Boeing, Martin-Marietta, General Dynamics, Rockwell International, and Lockheed, *Final Report of the Commercial Space Transportation Study*, NASA Headquarters, Washington, D.C., April 1994.

Bonometti, Robert, LTC, U.S. Army, *Perspectives on the Role of Satellites in the Information Infrastructure*, American Institute of Aeronautics and Astronautics/Utah State University Small Satellite Conference, Logan City, Utah, September 1993.

Broad, William J., "No Go for Satellite Sanctions Against Iran," *Science*, Vol. 208, May 16, 1980, pp. 685–686.

Builder, Carl, *The Icarus Syndrome*, Transaction Publishers, New Brunswick, 1994.

Builder, C. H., and T. W. Karasik, *Organizing, Training and Equipping the Air Force for Crises and Lesser Conflicts*, RAND, MR-626-AF, 1995.

Civilian Satellite Remote Sensing: A Strategic Approach, OTA-ISS-607, U.S. Government Printing Office, Washington, D.C., September 1994.

Cohen, William S., Secretary of Defense, *Report of the Quadrennial Defense Review*, Office of the Secretary of Defense, Washington, D.C., May 1997.

Command and Control of AFSPACE Forces: A White Paper Prepared to Articulate the Vision of COMAFSPACE, 14AF, Vandenberg AFB, final draft, 22 December 1997.

Cunniffe, Peter, *Misreading History: Government Intervention in the Development of Communication Satellites*, Master's thesis, Massachusetts Institute of Technology, Cambridge, Massachusetts, May 1990.

Davies, Merton E., and William R. Harris, *RAND's Role in the Evolution of Balloon and Satellite Observation Systems and Related U.S. Space Technology*, RAND, R-3692-RC, September 1988.

"DOD to Boost Civilian Satellite Use," *Space News*, December 5–11, 1994, p. 18.

Edelson, Burt, and Joseph Pelton, co-chairs, *Final Report of the NASA/NSF Panel on Satellite Communication Systems and Technology*, International Technology Research Institute, Washington, D.C., 1993.

Executive Office of the President, *Statement by the Principal Deputy Press Secretary to the President*, Office of the Press Secretary, Washington, D.C., September 16, 1983.

Executive Office of the President, *National Space Transportation Policy—NSTC 4*, Office of Science and Technology Policy, Washington, D.C., August 4, 1994a.

Executive Office of the President, *Remote Sensing Licensing and Exports—PD 23*, Office of the Press Secretary, Washington, D.C., March 10, 1994b.

Executive Office of the President, *Convergence of U.S. Polar-Orbiting Operational Environmental Satellite Systems—NSTC 2*, Office of the Press Secretary, Washington, D.C., May 10, 1994c.

Fogleman, Ronald R., GEN, USAF, "Need Better Mobility for War and Peace," *St. Louis Post-Dispatch*, May 5, 1994, p. 7B.

The Future of Remote Sensing from Space: Civilian Satellite Systems and Applications, OTA-ISC-558, U.S. Government Printing Office, Washington, D.C., July 1993.

Hagemeier, Col. Hal, et al., "Space Sensors Study," Air Force Materiel Command, Air Combat Command, and Air Force Space Command, Los Angeles Air Force Base, California, October 1995 (unpublished).

Hays, Peter, *Struggling Towards Space Doctrine: U.S. Military Space Plans, Programs, and Perceptions During the Cold War*, Ph.D. dissertation, Fletcher

School of Law and Diplomacy, Tufts University, Medford, Massachusetts, May 1994.

Hudson, Heather, *Communications Satellite*, Free Press, New York, 1990.

Johnson, Dana J., *The Evolution of U.S. Military Space Doctrine*, Ph.D. dissertation, University of Southern California, Los Angeles, California, 1987.

Johnson, Dana, J., *Trends in Space Control Capabilities and Ballistic Missile Threats: Implications for ASAT Arms Control*, RAND, P-7635, March 1990.

Johnson, Dana J., "Matching Requirements, Opportunities, and Resources: The Contribution of Space-Based Command and Control," Presentation, 62nd Military Operations Research Society Conference, United States Air Force Academy, Colorado Springs, Colorado, June 1994.

Johnson, Dana J., Katherine Poehlmann, and Richard Buenneke, *Space Roles, Missions, and Functions: Challenges of Organizational Reform*, RAND, PM-382-CRMAF, August 1995.

Johnson, Dana, Ken Reynolds, et al., "Integrating USAF Space Operations," RAND Project AIR FORCE briefing, Washington, D.C., December 1997.

Johnson, Nicholas L., "GLONASS Spacecraft," *GPS World*, November 1994.

Joint Vision 2010, Office of the Chairman of the Joint Chiefs of Staff, Joint Staff, Pentagon, Washington, D.C., April 1996.

Keegan, John, *The History of War*, Alfred Knopf, New York, 1993.

Killian, James R., *Sputniks, Scientists, and Eisenhower*, MIT Press, Cambridge, Massachusetts, 1977.

Lundberg, Olaf, INMARSAT Director-General, "The Future of Satellite Navigation and INMARSAT," Institute of Navigation Planning Conference, Salt Lake City, Utah, September 20, 1994.

Mack, Pamela, *Viewing the Earth: The Social Construction of Landsat*, MIT Press, Cambridge, Massachusetts, 1990.

Millot, Marc Dean, Roger Molander, and Peter Wilson, *The Day After . . . Study: Nuclear Proliferation in the Post-Cold War World. Volume I, Summary Report*, RAND, MR-266-AF, 1993.

Mission Revalidation Brief, USCINCSPACE to SECDEF, U.S. Space Command, Peterson AFB, Colorado, 7 November 1996.

Molander, R. C., A. S. Riddile, and P. Wilson, *Strategic Information Warfare: A New Face of War*, RAND, MR-661-OSD, 1996.

"NASA, U.S. Air Force Eye Technology Cooperation," *Space News*, December 15–21, 1997, p. 8.

Office of the Joint Chiefs of Staff, National Military Strategy, *Shape, Respond, Prepare Now: A Military Strategy for a New Era*, Washington, D.C., 1997.

Pace, Scott, *Factors Affecting the Potential for Commercial Space Launch Vehicles in South Africa*, U.S. Department of Commerce, Office of Space Commerce briefing, Washington, D.C., November 1992.

Pace, S., G. Frost, I. Lachow, D. Frelinger, D. Fossum, D. K. Wassem, and M. Pinto, *The Global Positioning System: Assessing National Policies*, RAND, MR-614-OSTP, 1995.

Pace, Scott, Kevin O'Connell, and Beth E. Lachman, *Using Intelligence Data for Environmental Needs: Balancing National Interests*, RAND, MR-799-CMS, 1996.

Poehlmann, Katherine, *Strategies for Achieving Commercial Space Sector Support to DoD Missions*, RAND, PM-384-CRMAF, August 1996.

Preston, Bob, *Plowshares and Power: The Military Use of Civil Space*, Institute for National Strategic Studies, National Defense University, 1994.

Report of the Advisory Committee on the Future of the U.S. Space Program, U.S. Government Printing Office, Washington, D.C., December 1990.

Sokolski, Henry, "South Africa Rethinks Its Space Goals," *Space News*, July 26–August 1, 1993.

"Space Launch Industry Faces Dramatic Changes," *Aviation Week and Space Technology*, December 16, 1996, p. 86.

Space Vest, *The State of Space Industry—1997 Outlook*, Space Publications, KPMG Peat Marwick, and Center for Wireless Telecommunications, Washington, D.C., 1997.

Steinberg, Gerald M., *The Legitimization of Reconnaissance Satellites: An Example of Informal Arms Control Negotiations*, Ph.D. dissertation, Cornell University, Ithaca, New York, 1981.

Studeman, William O., "The Space Business and National Security," *Aerospace America*, November 1994.

U.S. Congress, *Land Remote Sensing Policy Act of 1992*—P.L. 102-55.5.

U.S. Congress, *National Defense Authorization Act for 1993*, Report of the Senate Armed Services Committee, July 28, 1993.

U.S. Congress, Office of Technology Assessment, *Remotely-Sensed Data: Technology, Management, and Markets*, OTA-ISS-604, U.S. Government Printing Office, Washington, D.C., September 1994a.

U.S. Congress, *Civilian Satellite Remote Sensing: A Strategic Approach*, OTA-ISS-607, U.S. Government Printing Office, Washington, D.C., September 1994b.

U.S. Department of Defense, *Joint Staff Master Navigation Plan*, CJCSI 6130.01, May 20, 1994, National Technical Information Service, Washington, D.C., 1994.

U.S. Department of Defense, *Annual Report to the President*, March 1997.

U.S. Departments of Defense and Transportation, *Federal Radionavigation Plan*, DOT-VNTSC-RSPA-92-2/DOD-4650-5, National Technical Information Service, Washington, D.C., 1992.

U.S. Department of Transportation, Office of the Associate Administrator for Commercial Space Transportation, Federal Aviation Administration, *LEO Commercial Market Projections*, Washington, D.C., July 25, 1997.

U.S. House of Representatives, *Commercial Remote Sensing in the Post-Cold War Era*, joint hearings before the Committee on Science, Space, and Technology and the Permanent Select Committee on Intelligence, February 9, 1994.

USSPACE Vision Implementation Plan, U.S. Space Command, Peterson AFB, Colorado, 30 July 1997.

USSPACECOM Vision, U.S. Space Command, Peterson AFB, Colorado, 11 December 1996.

Vice President's Space Policy Advisory Board, *The Future of U.S. Space Launch Capabilities*, U.S. Government Printing Office, Washington, D.C., November 1992.

Weigley, Russell Frank, *The American Way of War*, Macmillan, New York, 1973.

The White House, Office of the Press Secretary, *Fact Sheet: U.S. National Space Policy*, November 16, 1989.

The White House, *National Space Policy*, National Science and Technology Council, Washington, D.C., September 19, 1996.

The White House, *A National Security Strategy for a New Century*, Washington, D.C., May 1997.

The White House, Office of the Press Secretary, *A National Security Strategy of Engagement and Enlargement*, n.d., p. 10.

Williams, Phil, "Transnational Criminal Organizations," *Survival*, Vol. 36, No. 1, Spring 1994.

Winnefeld, James A., and Dana J. Johnson, *Joint Air Operations: Pursuit of Unity in Command and Control, 1942–1991*, Naval Institute Press, Annapolis, Maryland, 1993.

Winnefeld, James A., Preston Niblack, and Dana J. Johnson, *A League of Airmen: U.S. Air Power in the Gulf War*, RAND, MR-343-AF, 1994.